Nitrogen in Agricultural Landscape

Editor
Lech Wojciech Szajdak
Institute for Agricultural and Forest Environment
Polish Academy of Sciences
Poznań, Poland

CRC Press is an imprint of the
Taylor & Francis Group, an **informa** business
A SCIENCE PUBLISHERS BOOK

Cover credit: Cover illustrations reproduced by kind courtesy of the editor.

First edition published 2021
by CRC Press
6000 Broken Sound Parkway NW, Suite 300, Boca Raton, FL 33487-2742

and by CRC Press
2 Park Square, Milton Park, Abingdon, Oxon, OX14 4RN

CRC Press is an imprint of Taylor & Francis Group, LLC

Library of Congress Cataloging-in-Publication Data
Names: Szajdak, Lech Wojciech, 1953- editor.
Title: Nitrogen in agricultural landscape / editor, Lech Wojciech Szajdak, Institute for Agricultural and Forest Environment, Polish Academy of Sciences, Poznań, Poland.
Description: First edition. | Boca Raton : CRC Press, 2021. | Includes bibliographical references and index.
Identifiers: LCCN 2020051515 | ISBN 9780367343156 (hardcover)
Subjects: LCSH: Nitrogen fertilizers.
Classification: LCC S651 .N576 2021 | DDC 631.8/4--dc23
LC record available at https://lccn.loc.gov/2020051515

ISBN: 978-0-367-34315-6 (hbk)
ISBN: 978-0-367-70740-8 (pbk)
ISBN: 978-0-429-35135-8 (ebk)

Typeset in Times New Roman
by Shubham Creation

Preface

Since the beginning of the 20th century the production of chemical and biochemical fertilizers and also pesticides and herbicides has increased significantly.

The chemistry of soils includes two parts, the chemistry of the various soils chemicals and biochemical and the chemistry of the soil processes and mechanisms.

Almost all nitrogen in surface soil horizons is in organic form (0-90%). Nevertheless, the chemical composition of nitrogen in organic soil fraction is not completely understood and little is known of the factors affecting the distribution of organic nitrogen forms in soils.

The availability of different forms of nitrogen in soil influences the net primary productivity and vegetation succession gradients in soils. The cycling of diverse N-containing substances in agriculture soils is still poor.

Inorganic and organic forms of nitrogen exist in the environment in various oxidation states and these substances are used by chemolithotrophic bacteria as a sole substratum of energy for growth. Understanding the function of various forms of nitrogen is necessary to discern their function in the environment.

Chemical processes and mechanisms in mineral and organic soils are largely estimated by reactions at the surfaces of the colloids. The colloidal complex of a soil plays a major role in supplying nutrients to plant grown thereon. Since humus is colloidal in nature its role in influencing soil fertility is considerable. As a colloid, humus is involved in three major functions: exchange capacity, buffer capacity and chelation of metals and organic compounds.

Chemical and biochemical compounds created as results of these processes influence of plant growth. The chemical structures of the main inorganic substances are now well known, but the many structures of organic compounds are still unknown.

Nitrogen enters into the soil cycle through N_2 oxidation by electrical discharge or dinitrogen assimilation or combustion processes. The plant cover uses all forms of nitrogen, organic and inorganic. Mineralisation, nitrification, denitrification and ammonification are major processes in nitrogen cycle. Various groups of organisms, bacteria, fungi and certain soil animals participate in nitrogen cycle and the product formed is generally, depending on the soils pH is uptake by roots of cultivated plants.

This book is related to the description of the processes of nitrogen transformation in the soil-plant system in mineral and organic soils of agricultural landscape.

It is about the distribution and equilibria between agricultural landscape compartments. It is about conversions, pathways thermodynamics and kinetics. An important aim of this book to help understanding of the basic distribution and chemical reaction processes which occur in the agricultural landscape.

The last explicated aim of this book is to present, in the concise form, the most important features relating to soil chemistry and crop yields.

Most chapters in the book should be of interest to the graduate students and practicing scientists. I also believe that the subject matter treated will be of interest to people outside agriculture and scientists in environmental chemistry.

This book would not have been possible without the significant cooperation of the contributing authors. I am indeed grateful to them for their excellent work during different stages of preparation of this book and the patience of my request, to the publisher for their encouragement.

Lech Wojciech Szajdak

Introduction

Lech Wojciech Szajdak

Institute for Agricultural and Forest Environment
Polish Academy of Sciences
Poznań, Poland

Nitrogen is one of the common elements for crops to achieve optimum yields. Organic and mineral forms of nitrogen are the common nutrients in demand by plants. The cycling of nitrogen in nature has been studied more extensively than that of any other nutrient. In soils of an agricultural landscape, the nitrogen occurs in soil biota and in the part of organic matter as a major constituent of peptizes, proteins, amino acids, alkaloids, amines, amides, nitrites, nitrates, ammonia, etc., and in the compounds from which these substances were synthesized.

The details of the biogeochemistry of the nitrogen cycle are available in the literature. Nitrogen occurs in the atmosphere; however, it must first be fixed from the atmosphere before it is suitable for uptake by plants.

The ability to fix N is connected among various species of the following groups: oxygenic and anoxygenic phototrophic bacteria, chemoautotrophic and chemoheterotrophic bacteria as well as aerobic and anaerobic bacteria.

In natural conditions, fixation of nitrogen occurs as a result of the complex microorganisms-enzymes, which are located in root nodules of certain higher plants (e.g., legumes). As a result of nitrogen fixation, organic and inorganic compounds of nitrogen and include in available forms, are commonly taken up by the different crops.

In soil, the following processes of the nitrogen cycle such as fixation, nitrification and denitrification belong to the most ecological steps in the nitrogen cycle.

Nitrogen fixation leads to the conversion of N_2 in the atmosphere into biologically-available ammonium. Nitrification transforms ammonium to nitrates under aerobic conditions. This inorganic form of nitrogen is important for cultivated plants, therefore nitrification is the factor of soil fertility and plant growth. Denitrification leads to the conversion of nitrate to N_2O and molecular dinitrogen (N_2).

Interaction between soil biota plays an important function in nutrients conversions and plant nutrient availability. The knowledge about the groups of soil microorganisms and organisms should shed some light on how microorganisms adapt to changing environmental conditions and how these changes are responsible for the productivity of soils and crop yields of cultivated plants.

Protozoa accelerates the mineralization of high molecular weight compounds of well-known and unknown structures and microbial immobilization of the nutrients. A major element regulating the rate, dynamics of the interaction between microorganisms and protozoa in soil is water. The drying of the soil strongly decreases the activity of soil protozoa.

The process of mineralization followed by nitrification and anthropogenic deposition will, in addition, combine and provide the inorganic forms of nitrogen—ammonium, nitrate, and nitrites in amounts and ratios which will depend upon local conditions of soil, climate and pollutant deposition.

More than half of the total nitrogen in soil may be unidentified, but in agricultural landscape organic forms of nitrogen, amino acids, proteins, peptizes, amino sugars alkaloids, peptidoglycans, phytohormones derivatives, etc., usually account for more than 95% of the identifiable organic nitrogen. Amino acids create the building blocks of proteins, making up the largest, very well functioning reservoir of nitrogen that is included in the organic compounds of most living organisms. Amino sugars and amino acids are included in microbial cell walls, and nucleic acids are located in living cells. Protein is a basic constituent of all life forms. During degradation, proteins are hydrolyzed to peptides by proteinases and peptidases. Mineralization of organic compounds of well known and unknown structures refers to the degradation of proteins, amino sugars, and nucleic acids with the formation of NH_4^+. Nitrogen in nonprotein structures is also included in the cell wall of bacterial and fungal chitin constitute.

The effect of returning crop residues to soil on nitrogen conversions are reasonably well understood. However, the most significant effects are on the relative rates of nitrogen immobilization and nitrogen mineralization. A well mature plant residues contain insufficient nitrogen to meet the need of the large microbial population, which arises in response to the addition of decomposable organic material, and net immobilization of inorganic N occurs. This condition is temporary and gives way to net mineralization after available C is depleted in a few days or weeks.

In view of the rapidly rising demand for agricultural products, use of the organic and mineral fertilizers and energy to satisfy a fast-growing world population, which is tending to become an essentially industrialized, raw-material-consuming society, it is appropriate to ascertain the natural resources of the earth which would provide the basic materials for sustainable development.

Nitrogen leaching is mainly connected with nitrates. The process depends on factors such as type and agricultural land use. Nitrogen leaching is restricted by shallow groundwater levels, owing to a higher capillary rise, increased supplementary superficial discharge (interflow and runoff), and increased denitrification. Under average drainage conditions, it is assumed that up to one-half of the leached N is lost in the subsoil by denitrification. In less aerobic to anaerobic groundwater, N usually occurs as ammonium.

The recognition of the feedback mechanism between the function of the countryside and the exploitation of goods or services needed by human society must take a first step in developing strategies for sustainable development of the agricultural landscape. In parallel with increasing recognition of agricultural landscape's basic processes such as organic matter cycling, energy fluxes, the mechanisms of their management, and the natural resources should be used but not for that they are exploited and degraded.

The organic matter and the concentration of nitrogen in organic compounds of most cultivated soils probably have now reached equilibrium under current management systems. Under some conservation tillage systems, the organic C and N contents in the soil appear to be increasing slightly but will not reach the levels contained in the soil in its native state. The nutrients supplied by the soil to a crop within a year very widely. For many high N-demanding crops, such as corn, the amount made available may range from 0 to 100 pounds or more per acre per year. More N will be available from soils high in organic matter, and in years following a legume in the rotation. If the characteristics of the soil and climate are known, experienced soil chemists should make reasonable estimates of available N.

The results of various studies presented in the subsequent chapters of this book clearly document this idea. Some of the results, showed more fully in the various chapters, are provided here to highlight the option for the management resources of nitrogen in agricultural soils and enable the development of guidelines for the implementation of environmentally friendly agricultural landscape management.

The book will help provide scientific basis required by students, researchers, teachers, agronomists, horticulturists, ecologists, environmentalists for laboratory practical classes worldwide and will be useful for scientists involved in the development of sustainable agroecosystems and contribute to a range of disciplines, including agriculture, horticulture, pomology, floriculture, biology, geography, landscape ecology, organic farming, biological control, environmental protection, and global change ecology.

Besides serving the researchers, the book will be also helpful for undergraduate and postgraduate students, soil scientists, biochemists, chemists, plant ecophysiologists, 'Natural Products' organic chemists, and other environmental scientists and specialists.

Agricultural scientists, representing many areas of specialization, have played a major role in studying the fate of nitrogen in the agricultural landscape, the effect of practices on soil, and the role of soils in the transformation processes involved. This publication is a compilation of symposium papers that were presented by scientists having a wide diversity of training and experience. Appreciation is expressed to all authors for preparing the manuscripts for publication.

Contents

Chapter 1

Nitrogen Content Under Blueberry Cultivation

Amalia Ioana Boţ[1*], Ioan Păcurar[1] and Lech Wojciech Szajdak[2]

[1]University of Agricultural Sciences and Veterinary Medicine, Cluj-Napoca, Romania.

[2]Institute for Agricultural and Forest Environment, Polish Academy of Sciences, Poznań, Poland.

INTRODUCTION

Nowadays there is a trend among the human population to consume food produced in ecological conditions that have a high nutrient content and supplies vitamins and minerals that are necessary for the good and healthy functioning of the human body. For this reason, an emphasis is placed on growing fruit and vegetables under environmentally friendly conditions with no chemical fertilizers; the crops grown on natural organic fertilizers are being increasingly appreciated.

Taking this into account, the question arises as to which crop culture can be set up under natural soil and climate conditions so that it can satisfy the current needs and requirements of the population at the same time. Therefore, in the light of the above, a study was carried out which aimed to both analyze the content of bio-chemical elements in the soil as well as the content of substances that have different benefits on the human body.

The target species for the research was highbush blueberry (*Vaccinium corymbosum* L.), which was chosen for several reasons. First of all, the growing cultivation trend of this species during recent years in Romania, which is why we

*For Correspondence: bot.amaliaioana@gmail.com.

have studied eight blueberry plantations located in the North-West Development Region of Romania. Second, the research done mainly in the medical field regarding the benefits of consuming blueberry fruits by the human body and the benefits arising from the antioxidant compounds found in the fruits was a major factor in deciding on the choice of this species for research.

The main objective of the study—which was discovered in the Ph.D. thesis titled "Assessment of Land Favorability from Northwest Region of Development for Blueberry Crop (*Vaccinium corymbosum*) in the Context of Sustainable Development of the Territory"—refers to the assessment of the pedological potential of certain soil types in the studied region for blueberry cultivation as well as the influence of certain soil compounds on the antioxidant capacity of blueberry fruits.

Listed above are a few of the reasons why this species stirs up an increasing interest among researchers in the agricultural and medical field. Moreover, another equally important aspect that has to be taken into consideration is the cultivation method of this species.

Organic or ecological farming has not only grown to meet people's needs for consuming organic products but also the conventional agriculture and industry is responsible for the climate change that we are currently facing. The gravity of the climate change situation in recent years has been noted at the UN Conference on Climate Change in Paris in 2015, reaching a unanimous agreement to "keep the average global temperature rise below 2°C, above pre-industrial levels and continue efforts to limit temperature rise to 1.5°C" as a long-term goal (unfccc.int/resource/docs/2015/cop21/eng/10a01.pdf).

On the basis of this agreement and the reports that have been drawn up over time by the institutions around the world, it was unanimously agreed that in order to avoid endangering long-term sustainable development of future generations, changes that take into account the present realities in industrial, technological and agricultural management are necessary (Boț 2017).

Blueberries are among the species that can be cultivated in an organic system, but above all, it can bring significant profit to farmers, which is why, besides its benefits to human health, this species is also a desirable way of farming nowadays.

Highbush blueberry (*Vaccinium corymbosum* L.) is a native species of North America and is also found in spontaneous flora of Romania. It prefers mountain areas, coniferous forests, soils with acidic pH and sunny exposure (Clapa 2006) but is increasingly gaining ground in cultivation for human consumption, which demonstrates awareness of the high economic potential and its antioxidant properties, as well as the beneficial effects of the human body, reinforced by the medical studies (Bunea et al. 2011, Forney et al. 2012, Gibson et al. 2013, Mehvesh et al. 2013, Kazim et al. 2015, Skrovankova et al. 2015, Cardeñosa et al. 2016).

Increased use of food rich in bioactive substances, vitamins and minerals, necessary for the optimal development of the body, awareness of the dangers that may arise from the effects of global climate change along with recent medical and agricultural research propels this shrub in the top of the varieties that simultaneously meet all the conditions listed (Boț 2017).

Another important aspect that needs to be taken into account when we want to motivate research on this shrub is also the increase in the number of blueberries cultivation and its world production. A study by Michalska and Lysiak in 2015 shows that the global production of blueberries has increased from 33,000 tons in 1965 to 420,000 tons in 2012. During 2009–2012, a FAO study showed that the global leader in blueberries production was the USA with an average production of 200,000 tons followed by Canada (93,000 tons) and Poland (10,600 tons).

Highbush blueberry (*Vaccinium corymbusum*) is a part of the genus *Vaccinium*, the *Ericacee* family, having its origin in North America. In Europe, highbush blueberry was grown especially in Holland and Germany since 1925, which then expanded to Austria, Denmark, England, Scotland, Switzerland, Italy and Poland. The first plantations in our country were set up in 1968 at the former Bilceşti Pomicol Research Center, Argeş County (Clapa 2006).

It prefers soils with pronounced acidity (pH 4.5–5.5) and peat bogs, soils located predominantly in the hilly and piedmont area of Romania, where the corresponding climatic conditions are met, respectively rainfall over 700 mm annually, and average annual temperatures of 7.8–8.5°C (Clapa 2006). Blueberry is suitable for acidic soils, rich in organic matter, well-drained, damp and with good luminosity characteristics. The land on which the crop is to be kept must be moist and free of stagnant water throughout the growing season and should be on sunny slopes with a slope of up to 20% without any sliding soil hazard. The edges of coniferous forests are the most desirable but not at an altitude of more than 800–900 meters.

Regarding the fertilizer application for blueberry crops, this is mainly done with organic fertilizers; the most often ones used are the manure and chemical fertilizers based on nitrogen, phosphorus and potassium, the ionic ammonia form (NH_4^+) of nitrogen being absorbed faster compared to the nitric form (NO_2^- and NO_3^-) which is absorbed with difficulty. At limiting pH values of 5.0–5.5, nitrifying bacteria are becoming more active, which convert available nitrogen into the soil in nitric form, hardly accessible to plants, while in more acidic soils rich in organic matter, the bacteria turn nitrogen in the ammonia form. Thus, on soils with pH above 5.0, it is recommended to use ammoniac nitrogen fertilizers (ammonium sulfate, ammonium nitrate or ammonium phosphate), and on more acidic soils with pH below 5.0, nitric fertilizers (ammonium nitrate or urea).

The nutrient requirement is crop-specific so that the pH of the nutrient substrate does not change considerably. On low organic matter soils, increasing doses of nitrogen are recommended starting from 10 kg/ha in the first year to 70 kg/ha in the seventh year. After the seventh year, an annual dose of 140 kg/ha is maintained. Nitrogen-based chemical fertilizers are applied fractionally in two equal doses applied to the flowering phenophase at the beginning and end of the process. The application of other fertilizers based on phosphorus, potassium and microelements must be correlated with their leaf and soil levels. For this, plant symptomatology should be considered (Mladin and Ancu 2014).

The most advantageous irrigation system of the blueberry culture is the drip system, which requires a low water consumption, and watering is done in the area of the nutrient bowl, thus eliminating the danger of diseases attacking leaves, fruits or stems. In the first years after planting, between 25 and 44 mm of water per week is required throughout the growing season. Young plants should receive 0.5l/day when temperatures are very high. The amount of water should increase by 0.5l/day for each year of the plant so that a plant in the sixth year after planting receives 3 l of water per day. It is also recommended to shut down the system for 12 hours to allow soil oxygenation.

Forest fruits are generally known as being rich in nutrients and containing a large number of vitamins, minerals and fiber. Blueberries are no exception to this rule; their polyphenolic content propel them into the top of the richest fruits in polyphenolic compounds and thus resulted in increased antioxidant properties. The terms underlying the description of the antioxidant capacity of blueberry fruits are (Krasovskaya et al. 2012):

- Oxidant – compound that oxidizes another compound;
- Antioxidant – compound which is capable of reducing the toxic effects of free radicals by neutralizing oxygen molecules, helping the body not to destroy its cells or compounds that have the ability to stop oxidation of other substances;
- Polyphenols – physiologically, they are essential for plant growth, for pigmentation, lignification and polinization; they are classified as antioxidants whose main activity is preventing the free radical formation and are found in the form of anthocyanins, flavonoids, quercitines, etc.;
- Anthocyans – natural pigments of the forest fruits, responsible for the appearance of blue, red or violet color; the highest concentration is found in the fruit skin (epicarp);
- Flavonoids – compounds with antioxidant properties, responsible for the yellow, orange and red colors, found in all fruits and vegetables.

Another definition given to compounds responsible for antioxidant properties appears in the study by Velioglu et al. (1998), where antioxidant substances are those compounds that inhibit or delay the oxidation of other molecules and which may be synthetic or natural. In general, synthetic antioxidants are compounds with different phenolic structures, while natural antioxidants can be phenolic compounds (tocopherols, flavonoids and phenolic acids), nitrogen compounds (alkaloids, chlorophyll derivatives, amino acids and amines) or carotenoids as well as ascorbic acid.

Many natural antioxidants, especially flavonoids, exhibit a wide range of biological effects, including antibacterial, antiviral, anti-inflammatory, anti-allergic and vasodilatory effects (Velioglu et al. 1998).

In Table 1.1, the chemical components of the cranberry fruits, their chemical structure, and the properties or action for each substance are identified.

Romanian authors (Diaconeasa et al. 2015) also note the importance of fruits and vegetables, rich in antioxidant compounds, consumption has long-term effects that are conditioned by the bioactivity of fruits and vegetables. Besides many bioactive compounds such as minerals, vitamins, sugars, organic acids and fibers, fruits and vegetables also contain phenolic compounds, such as flavonoids, tannins, stilbenes, phenolic acids and lignins.

Flavonoids include the anthocyanins, the largest natural pigments, which give the blue color to the fruits and show a very high antioxidant potential due to their chemical structure. The most common anthocyanins found in wild and cultivated blueberry fruits are cyanidins, dolphinidines, petunidines, paeonidines, pelargonidins and malvidins (Diaconeasa et al. 2015).

Table 1.1 Chemical structure and properties of blueberry compounds

Substance			***Chemical structure***	***Properties-effects on human health***
Polyphenols	Flavones	Kaempferol	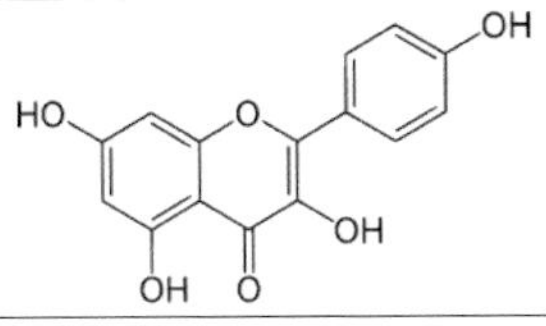	• inhibits the development of human malignant tumors by acting on glioblastoma cells • acts as an antioxidant, reducing oxidative stress
		Quercetin	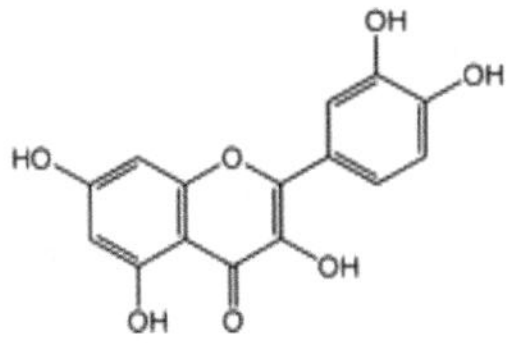	• antioxidant, fights against free radicals, reduces inflammation • natural antihistamine • increases immunity • improves skin health • natural supplement for cancer patients
		Myricetin	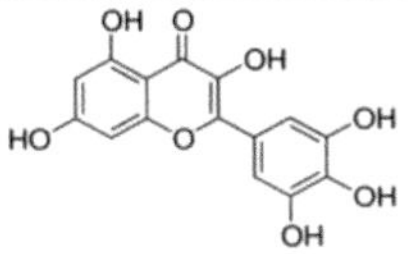	• strong anti-inflammatory and regenerative effect • anti-oxidant and anti-tumor properties
	Flavonoids		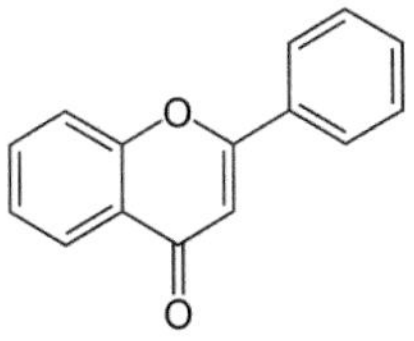	• reduces inflammation, anti-aging effect, reduces the risk of dementia, reduces the risk of heart disease and some cancers incidence, vasodilator effect • anti-bacterial, anti-viral, anti-inflammatory and anti-allergenic properties
	Procyanidins		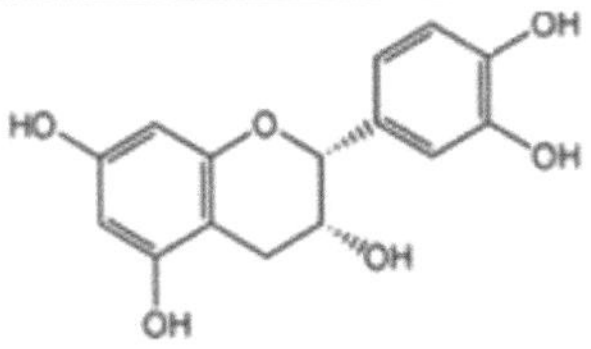	• reduces oxidative stress • anti-inflammatory and neuroprotective properties • reduces the risk of neurodegenerative and cardiovascular disease • reduces LDL cholesterol
	Phenolic acids		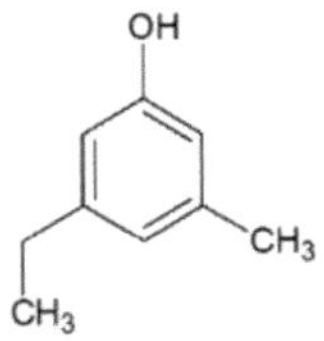	• anti-inflammatory and anti-diuretic properties • enhances immunity • helps common cold treatment • effect on tumor cells, especially in breast cancer • maintains skin health
	Stylbenic derivatives		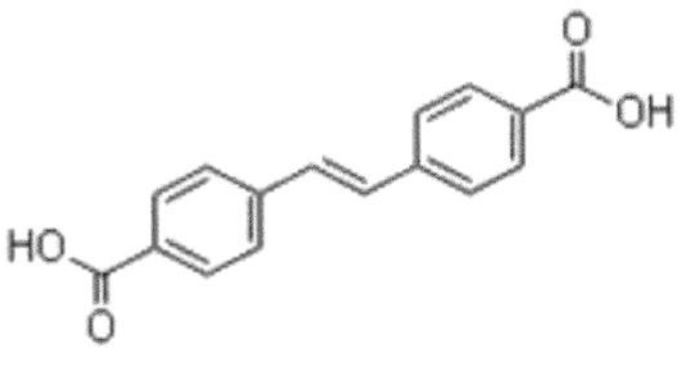	• antioxidant properties • beneficial effects on visual acuity

Anthocyanins	Malvidin	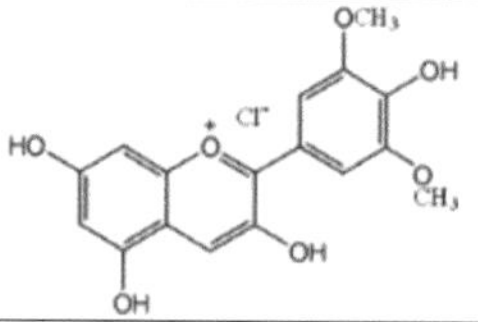	• antioxidant effect • cytotoxic effect on cancer cells, especially leukemia
	Delphinidin	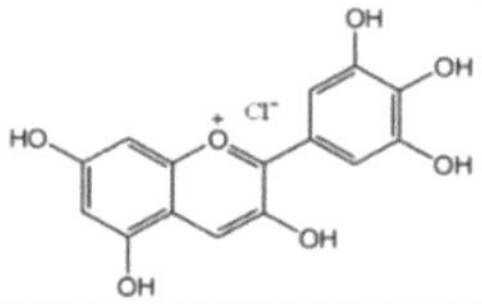	• antioxidant • regulates calcium homeostasis
	Petunidin	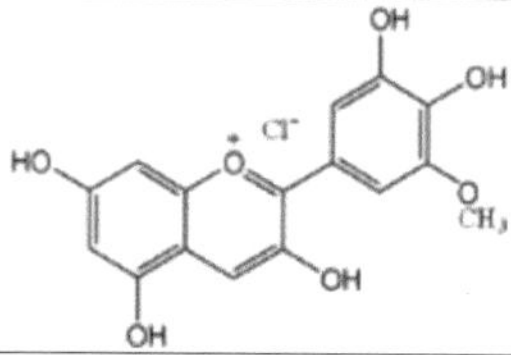	• helps in the treatment of hyperglycaemia, hyperinsulinemia and hyperelevantinemia • helps prevent obesity and diabetes
	Cyanindin	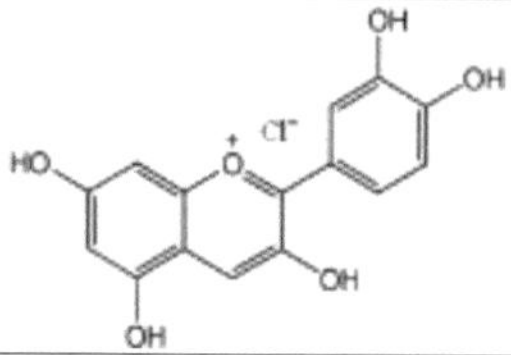	• antioxidant and free radical capture properties that can protect cells against oxidative stress and thus reduce the risk of cardiovascular disease and cancer cells
	Peonidin	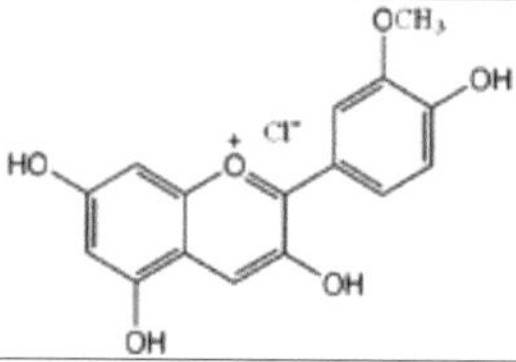	• an inhibitory effect on cancer cells, especially in the case of breast cancer metastasis • antioxidant effect
Chlorogenic acids		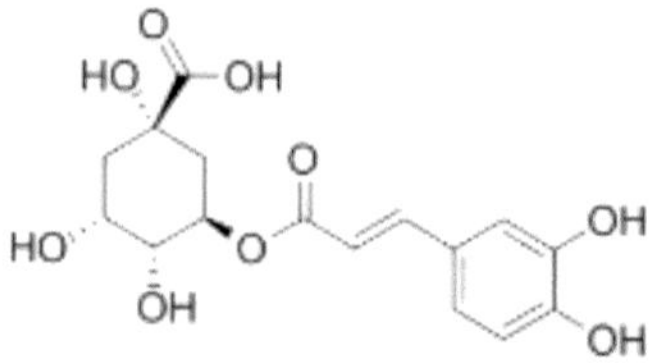	• inhibits the development of many forms of cancer • prevents and effectively treats diabetes and obesity • antioxidant potential • prevents cellular mutations • slows the aging process
Tannins		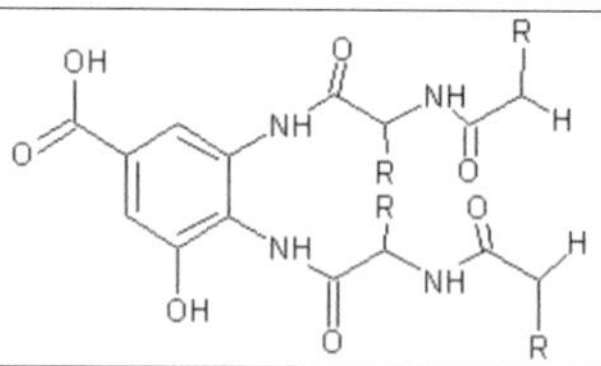	• excellent antioxidant • reduces cholesterol and improves the proportion of LDL and HDL • reduces blood pressure and cancer risk • stimulates the immune system • has antibacterial properties
Carotenoids		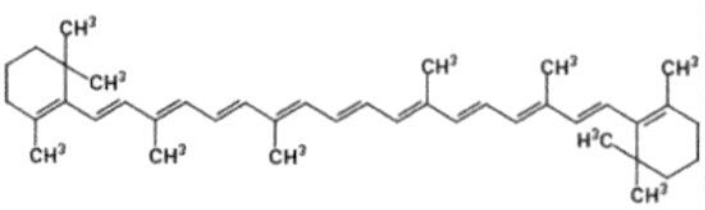	• has an antioxidant and skin-protective role, helping wound healing • used as anticancer substances, cell membrane stabilizing pigments, in the prevention and treatment of heart disease • gives strength to the immune system

Category		Structure	Effects
Folic acid		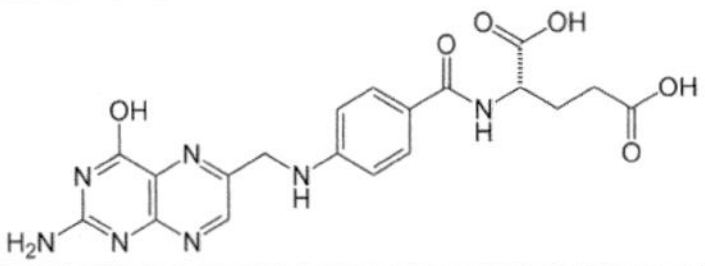	• nutrients for the pre-pregnancy period • treats various types of cancer • cohesion role for neurotransmitters
Vitamins	A	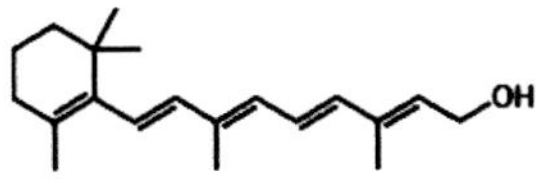	• maintains eye health, prevents inflammation and reduces the risk of cataracts • increases immunity • powerful antioxidant • prevents skin dryness
	C	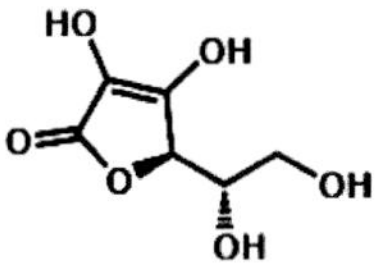	• reduces oxidative stress and free radicals • increases collagen production • strengthens the immune system • speeds up the skin healing • reduces the risk of cataracts • inhibits tumor growth • strong anti-inflammatory effect • participates in the synthesis of adrenal glands
	E	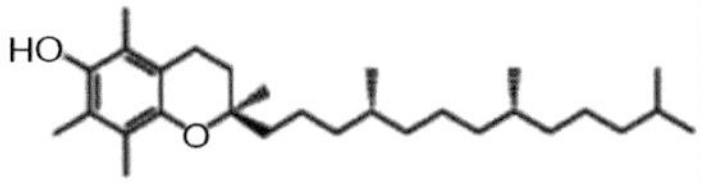	• antioxidant effect, reduces cell damage under free radicals • anti-cancer effect on prostate cancer • strengthens the immune system • protects against Alzheimer disease • decomposes fat-soluble toxins • prevents cardiovascular disease, atherosclerosis, cardiac ischemia, hypertension, myocardial infarction risk
Minerals	Calcium		• provides bone strength and muscle and nerve cells functioning • plays a role in the cell membrane permeability to ions, in the hormonal signals reception by cells and in enzymes activation
	Selenium		• powerful antiviral • helps thyroid correct functioning • hepatoprotective effect • biological antioxidant • stimulates antibody production, as well as fertility
	Zinc		• role in activating a large number of enzymes that are involved in protein synthesis, immunity and fight against colds • a powerful antioxidant effect on cancer cells • balances the hormones • promotes liver health, absorption and nutrient digestion

The richest sources of anthocyans, among fruits and vegetables, are the forest berries, such as blueberries (*Vaccinium corymbosum*), aronia (*Aronia melanocarpa*), *Sambucus nigra* and cranberry (*Vaccinium oxycoccos*). Numerous studies conducted so far on forest fruit extracts have demonstrated the link between forest fruits consumption and the decreased risk of developing different cancers (Stevenson and Scalzo 2012).

Blueberry fruits are among the most consumed forest fruits in Romania and are known for their benefits to human health. Recent studies have shown that a blueberry diet helps arterial structure by maintaining a healthy bloodstream due to LDL cholesterol oxidation, normal blood platelet aggregation and endothelial function improvement (Diaconeasa et al. 2014, 2015).

Therefore, following the information on the importance of blueberries, it can be stated that these fruits contain high concentrations of polyphenols, especially anthocyanins, phenolic acids, tannins, carotenoids, vitamin A, C and E, folic acid and minerals, such as Ca, Se and Zn. Blueberries contain high concentrations of anthocyanins and phenolic compounds with higher *in vitro* antioxidant activity compared to other fruits. Besides the beneficial effects on human health mentioned above, the consumption of blueberries helps in the treatment of biliary diseases, cough, tuberculosis and diabetes (Bunea et al. 2011).

INFLUENCE OF SOIL AND SOIL COMPOUNDS ON THE POLYPHENOLIC CONTENT OF BLUEBERRY FRUITS

In order to find the best solution for the economic growth attributed to blueberry cultivation, as well as the sustainable development of the areas where it is cultivated, by qualitative analysis of the land, farmers and researchers alike are looking to find the most suitable soils in terms of agro-pedo-climatic conditions and also the most profitable varieties in terms of crop yield given by the highest content of polyphenolic compounds.

Blueberry is a culture whose development in Romania, both in terms of fruit consumption and the establishment of new plantations, has increased significantly in recent years. The development of this agricultural sector has led to the search for the best varieties, in order to cope with competition, to meet the growing market demand and to meet high-quality standards. In this regard, referring to blueberry fruits, we can say that a high content of fruit polyphenols will automatically lead to an increase in crop economic profit by the increased quality of the obtained products.

Identifying the factors that influence the polyphenolic content of blueberry fruits is the first scientific step in determining the best varieties and crop technologies to support those who want to have such a plantation, both in terms of economic profit, and also in terms of the quality of the commercialized fruits.

Howard (2003) mentions that polyphenols are responsible for the beneficial effects exerted on human health through antioxidant activity and are influenced by genotype, season, culture location or soil conditions, fruit maturity and storage conditions.

Another study by Dragovic-Uzelac et al. (2010), identifies the same factors that influence the composition and polyphenolic content, namely species, seasons and agro-pedological conditions. In addition to these factors, there are also maturity periods of fruits, geographical origin of species, agricultural practices, storage conditions and processing methods (Łata et al. 2005).

Research on the factors influencing the polyphenolic content of blueberry fruits in our country was carried out by Bunea et al. (2011). They conducted an experiment in the North-West of Romania, consisting of two wild varieties of blueberry (*Vaccinium Myrillus*) and three cultivated varieties of *Vaccinium corymbosum*, Elliot, Bluecrop and Duke, and showed that wild varieties or spontaneously growing species in our flora have higher anthocyanin content than cultivated varieties. The same results were obtained by Koca and Karadeniz in an experiment in the Black Sea region of Turkey in 2009.

The antioxidant capacity of the fruits, their quality and the content of anthocyanins and phenolic compounds are influenced by the cultivation system and demonstrated that an organic cultivation system increases the concentration of these substances compared to the conventional cultivation system (Qi You et al. 2011).

Forney et al. (2012) conducted a study on the influence of maturation and ripening on fruit anthocyanins and demonstrated that the fruit ripening period (2–3 days) led to a rapid increase in pigments in the fruit, and implicitly, to an increase in the concentration of anthocyanins.

RESEARCH REGARDING THE NITROGEN CONTENT UNDER BLUEBERRY CULTIVATION

The study determined the land suitability for highbush blueberry and managed to identify which are the most suitable soil types for this crop, and at the same time succeeded in establishing, based on the nitrogen and its various forms content determination, the association between the soil biochemical compounds and blueberry fruits' bio compounds content as well as the influence of soil characteristics on the antioxidant properties of blueberry fruits.

Thus, in the Ph.D. thesis entitled "Assessment of Land Favorability from Northwest Region of Development for Blueberry Crop (*Vaccinium corymbosum*) in the Context of Sustainable Development of the Territory" (Boţ 2017), which refers to the assessment of the pedological potential of certain soil types in the studied region for the cultivation of blueberry as well as studying the influence of some soil compounds on the antioxidant capacity of blueberry fruits—eight plantations of blueberries were studied from the Northwest Region of Development from Romania (Figure 1.1)—while studying the areas and soil types with maximum favorability for the blueberry crop. Also, certain types of soil and fruit compounds have been determined in order to identify the influence that certain soil compounds may have on the blueberry compounds.

In order to achieve the results of the research, it was necessary to follow a methodology and to go through some steps. First, after the blueberry plantations were selected, the soil classes and types of soils were identified (Figure 1.1). The classifications of soils into classes and types were based on the Romanian Soil Taxonomy System, from 2012, developed by Florea N. and Munteanu from the Research Institute for Pedology and Agrochemistry (ICPA) Bucharest. In order to identify the type of soil from the eight blueberry plantations that were studied, soil profiles were made for each uncultivated plot of land near the plot of each of the eight blueberry crops, as well as morphological and physicochemical analyzes were conducted, according to Methodology of Pedological Studies, Volume I (Florea et al. 1987).

Analyzing the map of Romania's soils (1960), scale 1:50,000, it can be observed that within the studied territory, over the entire Northwest Development Region, predominates the class of luvisolles, followed closely by the cambisol and cernisol classes (Figure 1.1).

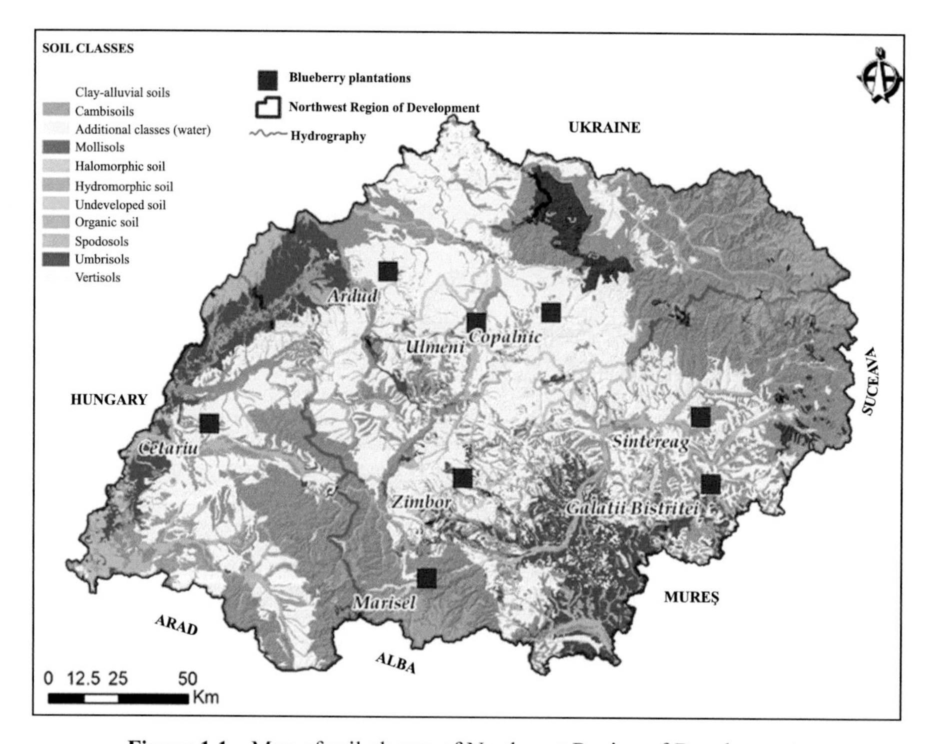

Figure 1.1 Map of soil classes of Northwest Region of Development.

In the figure above, besides the identification of the main soil classes in the Northwest Region of Development, which is the region that included the eight blueberry plantations that were studied, the soil types were also identified. This is of great importance because, based on these determinations, a more precise correlation between the physical and chemical properties of the soil and the biochemical compounds in the blueberry fruits can be done, that is correlation underlies the identification of the most suitable soils for this culture, both in terms of quantity, referring here to the generated crop yield and, qualitatively, mainly speaking about the improvement of the fruit quality, which also contributes to the maximization of the profit.

In addition to the identification of soil classes at the regional level, based on the Romanian Soil Map (1960) and the data provided by county OSPA offices, the soil types were identified in order to achieve the research goal. Thus, Figure 1.2 shows the map of the soil types at the level of the region, as well as the location of the parcels, in order to identify the type of soil corresponding to each plot.

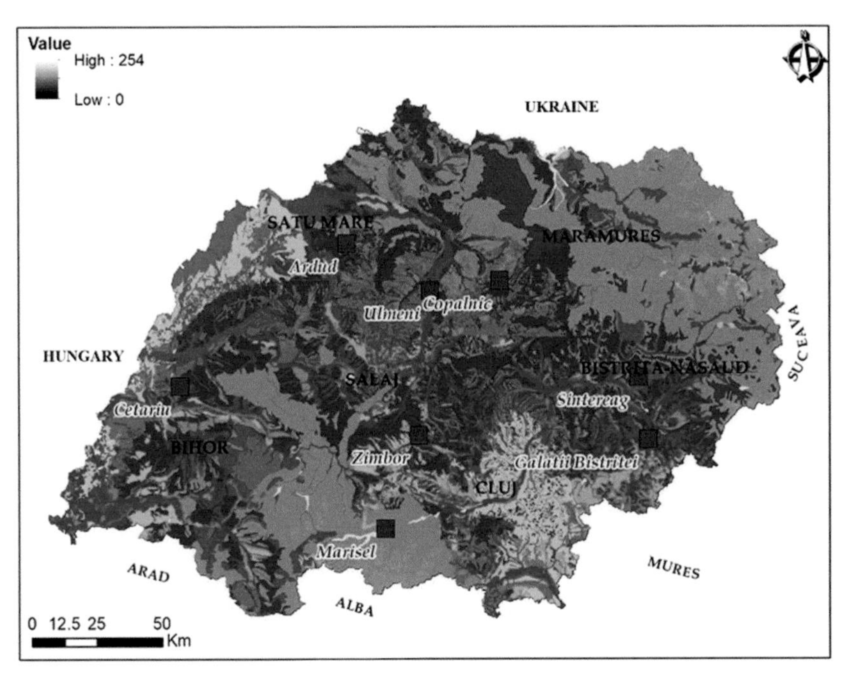

Figure 1.2 Map of soil type from Northwest Region of Development.

For the identification of the soil types, soil profiles were made at each of the eight blueberry plantations, the morphological description of the soil profiles that are part of the studied area being made based on the Romanian Soil Taxonomy System–SRTS, 2012. Soil classification at the type and subtype level was done after SRTS as well after World Reference Base for Soil Resources (WR–SR 1988) and the US Soil Classification System (USDA–Soil Taxonomy, 1999).

Research on the soil cover in the non-cultivated studied parcels under in order to determine the favorability of land for the blueberry crop was based on completing some field and laboratory steps necessary to determine the type of soil. Soil samples were taken for morphological characterization of soil profiles and were carried out on genetic horizons in the unmodified (natural) settlement for the characterization of physical, hydrophysical and micromorphological properties by using known metal cylinders (100 cm^2) at the soil moisture momentum (Boţ 2017).

The studied fields from Şintereag, Copalnic, Galaţii Bistriţei, Zimbor, Mărişel, Ulmeni, Ardud and Cetariu from the Northwest Region of Development

(Figure 1.2) present both the land plots cultivated with blueberries and the uncultivated plots. Based on the identification of soil types for the uncultivated parcels where the soil is in a natural habitat, the most common soil types were those in the class of luvisols, namely preluvosols, typical luvosol and albus luvosol followed by those in the class of cambisols.

The typical preluvosol, identified on the parcels were studied at Ardud and Zimbor; although usually suitable for most cultures with properties that impart high fertility, the basic character of parental rock is not favorable for the cultivation of blueberry. Typical and whitewashed luvosols, formed by more intense eluvial-iluvial processes than preluvosols, due to the poor parental material in basic elements, the humid and cool climate that favored altering and leaching processes as well as the presence of acidophilic natural vegetation (Filipov 2005) are among the most recommended soils for blueberry cultivation. Typical luvosols are present in the Copalnic and Cetariu blueberry crops, and the white one in the plot of Şintereag. Besides aluviosols, the eutricambosol soil type in the Ulmeni plot shows unfavorable characteristics for the blueberry crop because their parental material is rich in basic elements from basic magma rocks and sedimentary rocks that lead to the formation of humus type humus-calcic, imparting a neutral reaction to a low alkaline reaction (Păcurar 2000).

In the studied parcel from Mărişel, based on SRTS 2012, the identified districambosol has favorable characters for blueberry cultivation. The districambosols were formed on parental materials resulting from the disaggregation and alteration of acidic rocks, the formation processes consisting of alteration of the mineral part and acidic bioaccumulation (Păcurar 2010).

After identifying the soil types in the studied area, the soils with the highest natural favorability for the blueberry crop are those from Mărişel, where the districambosol was identified. Copalnic and Cetariu have also favorable conditions for the blueberry crop, both of which are determined by the presence of a typical luvosol and also the plot from Şintereag, where it was identified as soil type, the albic luvosol.

After determining the soil types, the field suitability for blueberry cultivation was identified. The physical and geographical characteristics of any territory will influence the vegetation factors and implicitly the agricultural productivity of the region through factors, such as altitude, slope, slope orientation, degree of fragmentation and so on. Several factors have been taken into account in determining the land favorability of the Northwest Region of Development, and the quality requirements for the crop have been taken into account. At the level of the whole region, climatic and geographic factors were taken into account for favorability, and at the level of each homogeneous ecological territory, namely the plots with the blueberry plantation, the pedological factor represented in the study of soil pH was added and generated favorability map for each studied plot (Boţ 2017).

The methodology used for determining the favorability map of the region and each homogeneous ecological territory (each of the eight blueberry crops) uses GIS technology in order to obtain a grid-based raster database, which was then used to identify land favorability, this methodology was being applied in similar studies (Bilaşco 2009, Halder 2013, Roşca et al. 2015, Moldovan et al. 2016).

Taking into account the requirements of a blueberry crop in order to determine the favorability for each of the parcels, we took into consideration the influence of several factors, such as the climatic factors (average annual temperature and annual mean precipitation), geographic factors (soil type, the inclination of slopes, altitude, land use), ecological factors (the bioactive period for blueberry) and soil factors (soil reaction) (Boț 2017). These factors represent raster databases that were used with G.I.S. program, which allowed statistical analysis, a reclassification of the factors and mathematical calculations and identification of the favorable and/ or restrictive areas for a certain type of culture by granting bonifications marks according to the influence of the analyzed environmental factors (Roşca 2014). For the classification of the factors by favorability classes, the norms of ICPA 1987 were taken into account as well as the requirements for cultivation: acidic soils with pH between 4.5 and 5.5, precipitation over 700 mm annually and the average annual temperature of 7.8–8.5°C (Clapa 2006), humus-rich soils, well-drained and humid but without stagnant water, sunny slopes (<200) without slip hazard and at altitudes between 800–900 m (Mladin and Ancu 2014).

Maps for each studied plot were digitized and assigned a raster grid data for each factor using Arc GIS 9.3. Each variable used was transformed into vector format, and the qualitative note of each favor class was assigned to it. Three classes of favorability have been established, to which qualitative marks have been awarded as follows: maximum favorability-3; medium favorability-2 and low favorability-1. Subsequently, the vector layers were combined by overlapping them in the GIS program, finally achieving favorable maps for each homogeneous ecological territory but also for the Northwest Region of Development. It should be noted that at the homogeneous ecological territory level, the pH factor was also used as the point values for each of the analyzed parcels, and at the regional level, this factor was eliminated in order to reduce the errors in the modeling program in the classification of favorability classes taking into account the climatic, geographic and ecological factors and the requirements of the blueberry crop (Boț 2017). For assigning a qualitative mark to the factor pH, it was taken into account the results obtained from determining the pH in each of the eight blueberry plantations in both the soil samples taken from the blueberry row and also from the soil samples taken from the plot in the natural state. Assigning a qualitative mark to the factor of pH, it was taken into account the optimum soil reaction for the blueberry crop, which was 4.5–5.5 (Clapa 2006).

The GIS technique used to identify land favorability for agricultural crops (Moldovan et al. 2016, Roşca 2014, Roşca et al. 2015) together with the applied methodology for identifying the soil types of the studied plots (Florea et al. 1987) highlighted the soil types which have the highest potential for blueberry cultivation. For each plot, the favorability (Figures 1.3 and 1.4) obtained from modeling which was based on modern analysis techniques that led to the development of raster databases is successfully correlated with the soil type of each plot. Thus, on parcels where luvosols and districambosols predominate, the highest favorability for blueberry that was identified (Figure 1.3a and b) was obtained for typical and albic luvosols, and medium favorability was obtained (Figure 1.3c and d) for districambosol and typical luvosol.

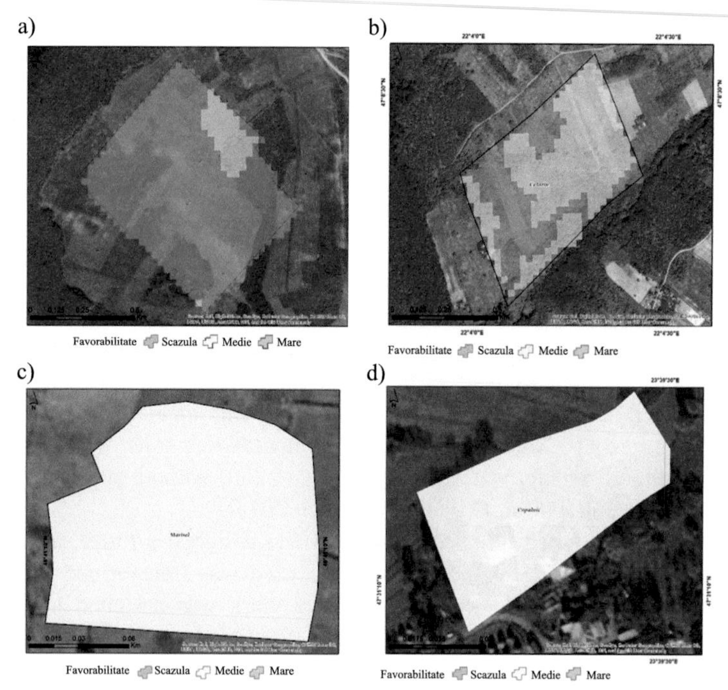

Figure 1.3 Land favorability for: a) Şintereag – high; b) Cetariu – high + medium; c) Mărişel – medium and d) Copalnic – medium.

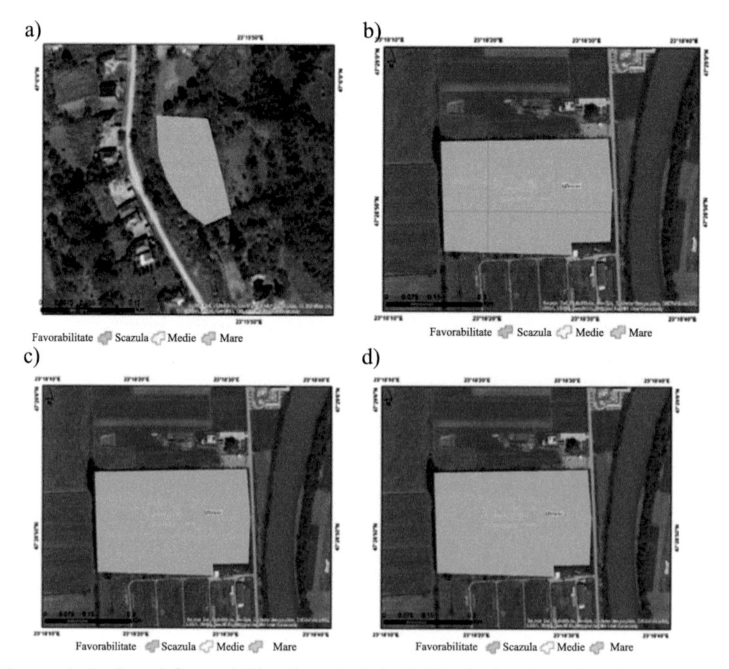

Figure 1.4 Land favorability for: a) Galaţii Bistriţei – low; b) Zimbor – low; c) Ulmeni – low; d) Ardud – low.

After applying the methodology for assessing the favorability of the studied homogeneous ecological territories (parcels), by overlapping all the factors used in the modeling process, the albic luvosol from Şintereag (Bistriţa-Năsăud county) was included in the maximum favorability class for blueberry cultivation followed by the typical luvosol soil of Cetariu (Bihor County) with a maximum and average favorability. In the plot of Mărişel (Cluj County), where the identified soil type is districambosol and the typical luvosol from Copalnic (Maramureş County), the land is in the middle favorability class for the cultivation of the blueberry (Boţ 2017).

The research below has been aimed at assessing the nutrient content and the various soil compounds. In order to collect the soil samples, it was chosen that the collection period to coincide with the vegetative rest period of blueberry. All the samples were taken from the blueberry plantations of the farmers from the counties of the Northwest Region of Development from Romania, where the study was conducted. In order to obtain relevant and comparative conclusions, two sets of samples were collected: a set of samples was taken from the plantation, harvesting the samples being made from the genetic horizon from the surface, and the second set of samples has been taken from the surface horizon of the soil profiles, which are located at a considerable distance from the blueberry rows, on the plots of land where the soil is in an unchanged settlement (natural state). Soil sampling was carried out according to Methodology of Pedological Studies, Volume I, ICPA Bucharest (Florea et al. 1987).

In order to form a relevant and concise conclusion, it was chosen to carry out a set of assays of soil samples taken: soil reaction, total organic carbon, dissolved organic carbon, ammonium and nitrate ions, total carbon and total nitrogen.

For the determination of *total nitrogen,* the Kjeldahl method was used. Solution nitrates (NO_3) were measured by the ion chromatographic method using the Schimadzu HIC-6A apparatus (Japan), and ammonium nitrogen (NH_4^+) was evaluated by the chromatographic method on the ion chromatograph Waters 1515 (USA) (Boţ 2017). Each soil sample was taken from a depth of 10–15 cm from the first horizon of the soil profiles of the non-cultivated parcels in the natural settlement, according to the methodology developed by ICPA, 1987.

Also, the C:N ratio was made to determine the soil mineralization process. Thus, both total and dissolved organic carbon were determined.

Total organic carbon was measured using the Shimadzu developed in Japan, Total Organic Carbon Analyzer (TOC 5050A) with Solid Sample Modules SSM-5000A. For determination, 50 mg of each sample was weighed, dried, grounded and sifted. Each sample was analyzed individually, the total organic carbon content was determined automatically by the apparatus. For analysis, three repetitions were used for each sample.

Dissolved organic carbon was measured with TOM 5050A developed by Shimadzu in Japan, using the liquid phase, for each sample was performed under three repetitions. The soil samples were heated to 100°C in deionized water for two hours under a reflux condenser, then cooled and centrifuged at 3,000 rpm for ten minutes. The extracts were filtered through 0.45 μm filter paper. A 5 ml of the filtrate were added to 50 ml volumetric flasks, which were then filled with double distilled water up to the cut-off limit and analyzed using TOC 5050A. Calculation of the content (mg/100 g of soil) is (1):

$$= C \times V \times 10/m \times V_{extract} \quad (1)$$

where: C–concentration; V–volume of the flask (50 ml); m–mass of soil; $V_{extract}$– extract volume; 10–conversion factor.

The results obtained for the soil samples taken from the affine plantations from the naturally occurring soil are found in Table 1.2

Table 1.2 Biochemical properties of the studied soils in a natural state

Soil type	***Luvosol albic (Şintereag)***	***Aluviosol (Galații bistriței)***	***Luvosol typic (Cetariu)***	***Districambosol (Mărișel)***	***Eutricambosol (Ulmeni)***	***Luvosol typic (Copalnic)***	***Preluvosol (Ardud)***	***Preluvosol (Zimbor)***
Horizon (10 cm)				Ao				
Sample/ Analyze	1	2	3	4	5	6	7	8
pH	5.30	7.20	6.20	4.90	7.00	5.40	7.20	6.90
N_{total} (g/kg)	2.30 ±0.24	1.19 ±0.05	1.08 ±0.05	2.43 ±0.49	1.44 ±0.12	1.27 ±0.25	1.70 ±0.12	0.96 ±0.02
NO_3^- (mg/kg)	7.86 ±0.44	9.89 ±0.42	4.97 ±0.42	8.40 ±0.00	4.75 ±0.34	9.66 ±0.37	2.01 ±0.13	4.39 ±0.23
NH_4^+ (mg/kg)	2.57 ±0.09	6.49 ±0.54	3.73 ±0.54	6.57 ±0.09	5.61 ±0.53	3.96 ±0.35	8.53 ±0.53	8.67 ±0.51
TOC (g/kg)	37.67 ±0.98	19.03 ±0.29	13.12 ±1.29	52.06 ±1.62	21.23 ±0.29	19.23 ±0.20	27.30 ±0.29	12.86 ±1.02
DOC (g/kg)	1.88 ±0.23	1.65 ±0.14	1.11 ±0.14	4.36 ±0.12	1.21 ±0.14	1.10 ±0.04	1.69 ±0.14	1.35 ±0.20
C:N	14	13	10	18	12	12	13	11

*The data in the table represent the mean values of the parameters analyzed±SD

In order to interpret the results obtained from the determination of the total nitrogen content for each sample, the value obtained was converted from g/kg to percent as follows: Zimbor typical preluvosol was classified as having a very low content of N_{total} (0.096%), the typical luvosol from Cetariu (0.108%), the aluviosl from Galații Bistriței (0.119%) and the typical luvosol from Copalnic (0.127%). With a medium content of N_{total}, soil types were identified as follows: eutricambosol from Ulmeni (0.144%), typical preluvosol from Ardud (0.170%), albic luvosol from Şintereag (0.230%) and districambosol from Mărişel (0.243%).

The most favorable pH values obtained from the samples taken from the uncultivated plots were obtained from the luvosol albic from Şintereag (5.3), the districambosol from Mărişel (4.9) and the typical luvosol from Copalnic (5.4). Following the methodology for the assessment of the land favorability for the blueberry crop, the parcels from Şintereag, Mărişel and Copalnic have high and medium favorability for these uncultivated parcels, the values obtained from the pH determinations coming to strengthen the results obtained after the assessment of land favorability for these fields (Boț 2017).

Table 1.3 Biochemical properties of the studied soils in a natural state

Soil type	***Luvosol albic (Șintereag)***	***Aluviosol (Galații bistriței)***	***Luvosol typic (Cetariu)***	***Districambosol (Mărișel)***	***Eutricambosol (Ulmeni)***	***Luvosol typic (Copalnic)***	***Preluvosol (Ardud)***	***Preluvosol (Zimbor)***
Horizon (10 cm)	Ao							
Sample/ Analyze	1	2	3	4	5	6	7	8
pH	5.30	7.20	6.20	4.90	7.00	5.40	7.20	6.90
N_{total} (g/kg)	2.30 ±0.24	1.19 ±0.05	1.08 ±0.05	2.43 ±0.49	1.44 ±0.12	1.27 ±0.25	1.70 ±0.12	0.96 ±0.02
NO_3^- (mg/kg)	7.86 ±0.44	9.89 ±0.42	4.97 ±0.42	8.40 ±0.00	4.75 ±0.34	9.66 ±0.37	2.01 ±0.13	4.39 ±0.23
NH_4^+ (mg/kg)	2.57 ±0.09	6.49 ±0.54	3.73 ±0.54	6.57 ±0.09	5.61 ±0.53	3.96 ±0.35	8.53 ±0.53	8.67 ±0.51
TOC (g/kg)	37.67 ±0.98	19.03 ±0.29	13.12 ±1.29	52.06 ±1.62	21.23 ±0.29	19.23 ±0.20	27.30 ±0.29	12.86 ±1.02
DOC (g/kg)	1.88 ±0.23	1.65 ±0.14	1.11 ±0.14	4.36 ±0.12	1.21 ±0.14	1.10 ±0.04	1.69 ±0.14	1.35 ±0.20
C:N	14	13	10	18	12	12	13	11

*The data in the table represent the mean values of the parameters analyzed±SD

Interpretation of the results obtained is based on the rules laid down by ICPA, 1987 (Borlan et al. 1973, 1981) for the total nitrogen content of the uncultivated soils.

Table 1.4 Evaluating the total nitrogen content (according to ICPA, 1987)

Assessing the level of nitrogen	*%N_{total}*
Very low	<0.100
Low	0.100–0.140
Medium	0.141–0.270
High	0.271–0.600
Very high	>0.600

Based on the DUNCAN test, multiple comparisons were made between the eight locations with blueberry plantations for the same chemical compounds analyzed, namely the total nitrogen and the amount of nitrate and ammonium, to compare the differences in compound concentrations. First, the determinations obtained on the organic and mineral forms of nitrogen from all analyzed soil types (Figure 1.5) were compared from the samples from the uncultivated plots.

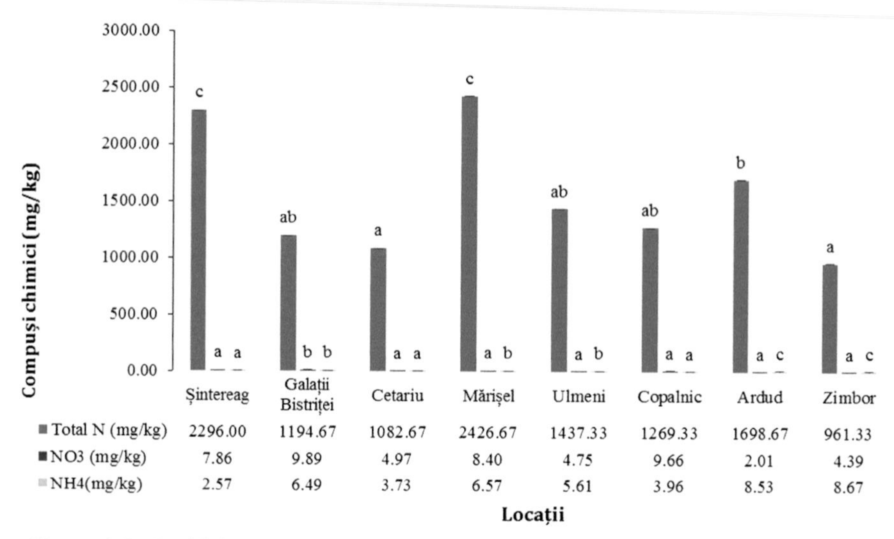

Figure 1.5 Multiple comparison analysis for nitrate forms (DUNCAN Test, $p = 0.05$).

From a statistical point of view between the soil type districambosol (Mărișel) and albic luvosol (Șintereag), there are no significant differences in the content of N_{total}; here the highest values are recorded. Significantly different values are between the soil of Ardud (b), (preluvosol) and the samples collected from Șintereag (c) and Mărișel (c) as well as between the soil of Zimbor (a) (preluvosol), the last one having lower values of total nitrogen. There are no significant differences between the soils from Galați Bistrita (aluviosol), Ulmeni (eutricambosol) and Copalnic (typical luvosol), also referring to the total nitrogen concentration in the uncultivated soil.

Significantly positive relationships are observed for dissolved and total organic carbon and C:N ratio but also between total nitrogen and C:N ratio and their increases in soil increasing the ratio.

After this comparison, multiple comparisons were made based on the same statistical method, but this time all eight samples were compared between them in terms of the content of the mineral and organic forms of nitrogen and also the total and dissolved organic (Figure 1.6) chemical compounds with significant value for soils that are not cultivated.

Concerning the comparison of the results for the carbon forms contained in the different analyzed soils, significant statistical differences exist between districambosol (Mărișel) and the rest of the analyzed samples where the total organic carbon has the highest value, which confirmed the average humus content that was determined for this type of soil following the application of morphological and physicochemical analyzes to describe the soil types identified in the study area. Referring to the total organic carbon content, we also note that there are no significant differences between aluviosol (Galații Bistriței), eutricambosol (Ulmeni) and typical luvosol (Copalnic). For dissolved organic carbon, following the DUNCAN test for multiple comparisons, between districambosol and the rest of the analyzed samples, there are significant statistical differences.

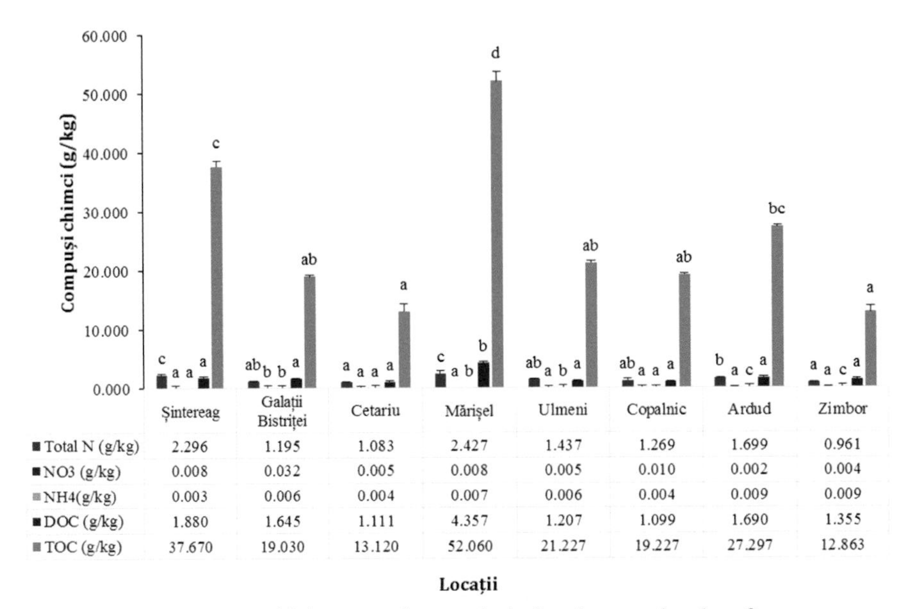

	Șintereag	Galații Bistriței	Cetariu	Mărișel	Ulmeni	Copalnic	Ardud	Zimbor
Total N (g/kg)	2.296	1.195	1.083	2.427	1.437	1.269	1.699	0.961
NO3 (g/kg)	0.008	0.032	0.005	0.008	0.005	0.010	0.002	0.004
NH4(g/kg)	0.003	0.006	0.004	0.007	0.006	0.004	0.009	0.009
DOC (g/kg)	1.880	1.645	1.111	4.357	1.207	1.099	1.690	1.355
TOC (g/kg)	37.670	19.030	13.120	52.060	21.227	19.227	27.297	12.863

Figure 1.6 Multiple comparison analysis for nitrate and carbon forms (DUNCAN Test, $p = 0.05$).

The results obtained for soil samples taken from the blueberry plantations from the billon are shown in Figure 1.6. For each analyzed soil sample, three repetitions of the determinations were performed and the values obtained represent the average (Table 1.5), the analyzes being carried out according to the methodology presented previously.

The results obtained by collecting the soil samples from each blueberry plantation reflects how the blueberry crop technology can influence the variability of some biochemical compounds analyzed in the soil. For each parcel, the farmers used the same cultivation method, namely planting the blueberry in a mixture of soil, sawdust and acid peat in varying proportions, depending on the soil pH before setting up the plantation.

So, given that crop technology remains unchanged and the variety of blueberry used is the same for all plantations (the bluecrop season variety), the only factor that differs from one plot to another being the soil type, it was wanted to identify possible changes to the soil properties that may occur after the establishment of such a crop, taking into account the blueberry cultivation technology, in order to determine what kind of soil best performs after cultivation in terms of soil compounds.

The evaluation of the total nitrogen content, performed according to the ICPA Bucharest norms, 1987, based on the studies conducted by Borlan et al. (1973 and 1981), fits the results obtained with the Cetariu sample from typical luvosol to a very low level of total nitrogen, the eutricambosol from Ulmeni to a small level and the rest of the analyzed samples had a mean content of total nitrogen. Following the establishment of a blueberry crop on these soils, it is noted that total nitrogen remained in the same concentrations after planting for albic luvosol (Șintereag), districambosol (Mărișel) and preluvosol (Ardud), having a total average nitrogen content.

Table 1.5 Biochemical properties of the studied soils (planting rows)

Soil type	***Luvosol albic (Şintereag)***	***Aluviosol (Galaţii bistriţei)***	***Luvosol typic (Cetariu)***	***Districambosol (Mărişel)***	***Eutricambosol (Ulmeni)***	***Luvosol typic (Copalnic)***	***Preluvosol (Ardud)***	***Preluvosol (Zimbor)***
Horizon (10 cm)	Ao							
Sample/ Analyze	1	2	3	4	5	6	7	8
pH	5.00	6.8	5.40	4.90	6.10	5.00	6.60	6.40
N_{total} (g/kg)	1.68 ±0.24	1.75 ±0.46	0.90 ±0.24	1.75 ±0.09	1.34 ±0.24	1.47 ±0.05	1.68 ±0.28	1.66 ±0.12
NO_3^- (mg/kg)	14.67 ±0.34	31.67 ±0.97	3.66 ±0.13	31.67 ±1.23	7.15 ±0.53	30.30 ±0.42	14.67 ±0.66	24.47 ±0.97
NH_4^+ (mg/kg)	1.48 ±0.09	9.37 ±0.68	3.38 ±0.09	9.37 ±0.29	2.18 ±0.09	6.60 ±0.54	1.48 ±0.28	12.07 ±0.53
TOC (g/kg)	28.36 ±0.98	26.40 ±1.21	9.32 ±0.78	26.40 ±0.77	19.85 ±0.98	21.83 ±1.29	28.36 ±0.28	51.64 ±0.29
DOC (g/kg)	1.25 ±0.23	2.06 ±0.35	0.80 ±0.23	2.06 ±0.06	1.19 ±0.23	1.22 ±0.14	1.25 ±0.28	3.06 ±0.14
C:N	13	12	8	12	12	12	13	31

*The data in the table represent the mean values of the parameters analyzed±SD

Regarding the Ulmeni eutricambosol type of soil and typical luvosol from Cetariu, the total nitrogen content drops from medium to low, from small to very small, indicating that the plants use total nitrogen in the soil under its mineral forms of nitrate and ammonium. This is confirmed by the fact that on all these soils the values of $N–NO_3^-$ and $N–NH_4^+$ are the lowest of all the analyzed samples (3.66 mg/kg $N–NO_3^-$ at Cetariu and 7.15 mg/kg $N–NO_3^-$ to Ulmeni), which is also supported by the value of the C:N ratio determined below 15:1, favoring the soil mineralization process.

In order to provide the data from the statistical point of view, the most important biochemical compounds from soil using Duncan's test was performed a number of multiple comparisons between all soil types in this research.

In the first step, the concentrations obtained in terms of the total nitrogen content but also of its mineral forms (Figure 1.7) were compared, which are used for plant nutrition.

From the graphic of multiple comparisons of organic and mineral nitrogen forms, it can be noticed that there are no statistically significant differences between the soil types of Şintereag (albic luvosol), Galaţii Bistriţei (aluviosol), Mărişel (districambosol) and Zimbor (preluvosol), in terms of total nitrogen concentrations, in these soils being recorded the highest values. For the same compound analyzed, distinct significant differences occur in the typical luvosol from Cetariu (a), compared to all other soil types, the value determined for this type of soil being the lowest.

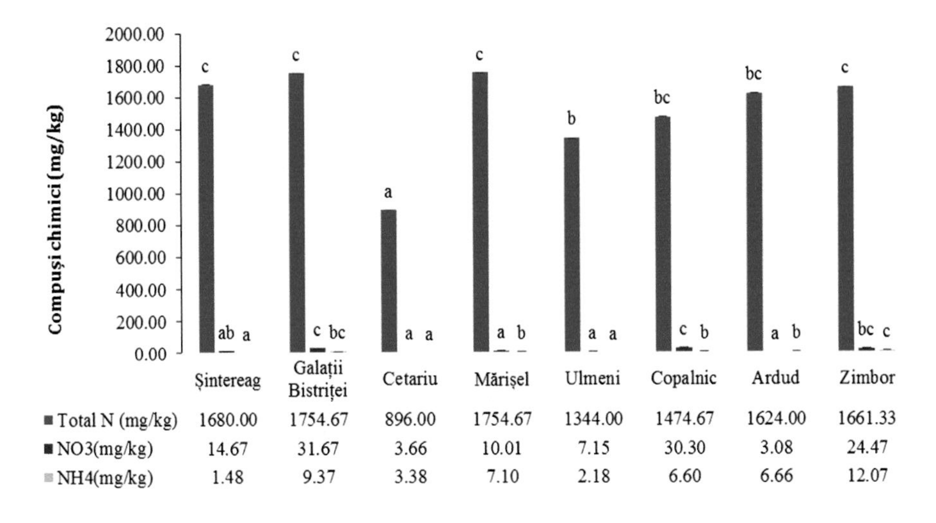

	Șintereag	Galații Bistriței	Cetariu	Mărișel	Ulmeni	Copalnic	Ardud	Zimbor
Total N (mg/kg)	1680.00	1754.67	896.00	1754.67	1344.00	1474.67	1624.00	1661.33
NO3(mg/kg)	14.67	31.67	3.66	10.01	7.15	30.30	3.08	24.47
NH4(mg/kg)	1.48	9.37	3.38	7.10	2.18	6.60	6.66	12.07

Figure 1.7 Multiple comparison analysis for nitrate forms (DUNCAN Test, $p = 0.05$).

Concerning the ammonium content, significant differences occur between the typical luvosol (Zimbor) and the rest of the soil types, less of the albic luvosol (Galații Bistriței), between the two is significant statistical differences.

In order to compare the statistical results with the carbon content of soils cultivated with blueberries, Figure 1.8 shows the results obtained following the DUNCAN test for multiple comparisons.

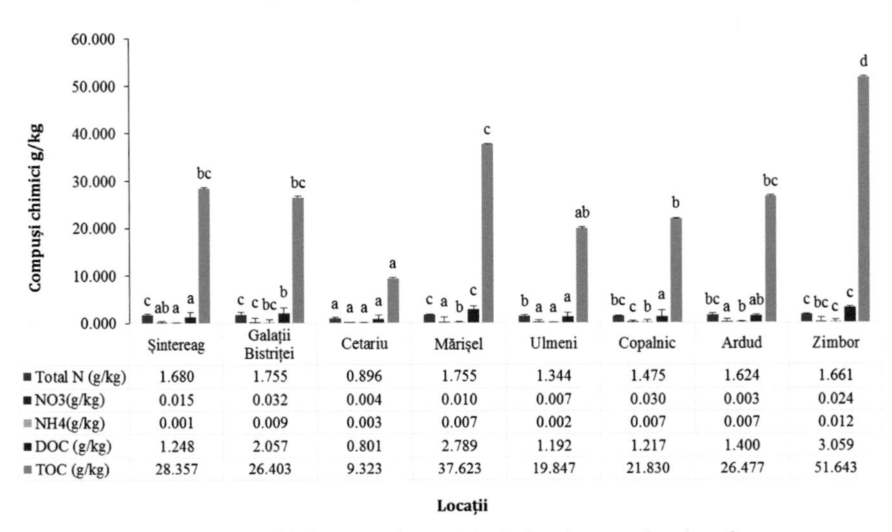

	Șintereag	Galații Bistriței	Cetariu	Mărișel	Ulmeni	Copalnic	Ardud	Zimbor
Total N (g/kg)	1.680	1.755	0.896	1.755	1.344	1.475	1.624	1.661
NO3(g/kg)	0.015	0.032	0.004	0.010	0.007	0.030	0.003	0.024
NH4(g/kg)	0.001	0.009	0.003	0.007	0.002	0.007	0.007	0.012
DOC (g/kg)	1.248	2.057	0.801	2.789	1.192	1.217	1.400	3.059
TOC (g/kg)	28.357	26.403	9.323	37.623	19.847	21.830	26.477	51.643

Figure 1.8 Multiple comparison analysis for nitrate and carbon forms (DUNCAN Test, $p = 0.05$).

The DUNCAN test applied for statistical validation for multiple comparisons between the values of the various compounds analyzed in the soil confirms the results obtained for the total organic carbon determined from Zimbor preluvosol (c) as the highest recorded value, distinctly significant from the other analyzed samples. Also, related to the total organic carbon in soils cultivated with blueberries, there are no significant differences between the soils of Şintereag (albic luvosol), Galaţii Bistriţei (aluviosol) and Ardud (preluvosol) or among soil types from Copalnic (typical luvosol) and Ulmeni (eutricambosol). The soil sample analyzed from Cetariu, in terms of total organic carbon content, is significantly different from all the other analyzed samples with the lowest value.

Based on the soil analyzes performed to determine the content of some soil compounds respectively to characterize the different soil properties as well as the influence of a particular type of technology applied for the cultivation of blueberry, it was possible to identify soil characteristics that change in after cultivation.

Analyzes carried out on soil samples taken from field plots where the soil is in a natural state have made it possible to identify the type of soil that has the best properties and which has the best qualitative properties in the natural state suitable for blueberry cultivation. These determinations can be corroborated with the study on the assessment of land favorability for this type of culture. In order to determine the areas where the blueberries present the best qualitative properties but also to identify the land suitable for this crop, other studies have been carried out on the favorability of the land for the blueberry crop and the influence of different soil properties on the concentration of the bioactive compounds from blueberry fruits.

In order to achieve the suitability of the field for blueberries and to carry out analyzes of the biochemical content of fruits as well as their antioxidant capacity, the present study uses the bluecrop variety (Figure 1.9) of the high-bush (*Vaccinium corymbosum*) category.

Figure 1.9 *Vaccinium corymbosum*, Bluecrop variety. *Source*: original.

Among the most important bioactive compounds in blueberry fruits are polyphenols and flavonoids, which have many benefits for human health. From the category of total polyphenols, along with flavonoids, anthocyanins, bioactive flavonoid compounds are also beneficial to many chronic diseases. Su and Chien (2007), quoted by Routray and Orsat in 2011, show that blueberry fruits have among the highest concentrations of anthocyanins and other polyphenols and exhibit a high antioxidant capacity.

All the analyzed samples came from the fruits harvested from the eight plantations of the Northwest Region of Development from their maximum period of ripening, the variety of seasonal blueberries, Bluecrop, high-bush spp. *Vaccinium corymbosum*.

The *total polyphenolic content* was determined using the Folin Ciocalteu Colorimetric Protocol (FCR) with some minor modifications from Singleton et al. (1999). The Folin-Ciocalteu method is a fast and light spectrophotometric screening method used to express the total content of all phenols, including flavonoids, procyanidins, anthocyanins and non-flavonoids. The working method consists of making a mixture consisting of an aliquot part of the extract (25 μl sample), a calibration standard of gallic acid and a blank sample in water were placed in separate plastic cuvettes. Then, 1.8 ml of distilled water was added to each well followed by 120 μl of Folin-Ciocalteu reagent, the samples were mixed and incubated for five minutes. After incubation, 340 μl Na_2CO_3 (7.5% Na_2CO_3 in water) was added to create basic conditions (pH ~ 10). The obtained mixture was incubated for 90 min at room temperature, and then the absorbance was read at 750 nm using a microplate reader (BioTek Instruments, Winooski, VT). The absorbance of the blank sample was subtracted from all readings and a calibration curve was created from the standards. The number of total polyphenols was expressed in mg of gallic acid/100 g of fresh material. Analyzes were performed in three rehearsals.

The *total flavonoid content* in the anthocyanin-rich fractions of blueberries was determined by a colorimetric method used by other authors (Kim et al. 2003, Zhishen et al. 1999). The obtained extracts were diluted with distilled water to a final volume of 5 ml and then 300 μl of 5% $NaNO_2$ was added. After 5 minutes, 300 μl of 10% $AlCl_3$ was added and after 6 minutes 2 ml of 1N NaOH. The solution was well mixed and the absorbance was read at 720 nm using a spectrophotometer (JASCO V–630 series, International Co., Ltd., Japan) and the total amount of flavonoids was expressed in mg quercetin equivalents/100 g of fresh material. Each determination was made in three rehearsals.

The neutralization effect of ABTS radical is a method that is based on ABTS + neutralizing ability by antioxidants (anthocyanins) compared to a standard antioxidant (vitamin E analog, Trolox). This analysis was performed according to the procedure described by Arnao et al. (2001) adapted to the 96-well plate. The green-blue solution of ABTS was produced by the reaction between the 7 mM ABTS aqueous solution and 2.45 mM potassium persulphate in the dark at room temperature for 12–16 hours before use. The working ABTS solution was obtained by diluting the stock solution with ethanol to obtain an absorbance of 0.70 ± 0.02 AU at 734 nm. A 20 μl of Trolox or assay of different concentrations were mixed with 170 μl ABTS solution and the absorbance of the samples was read after 6 minutes incubation at room temperature in the dark with a microplate reader (BioTek Instruments, Winooski, VT). The results were expressed as μmol of Trolox/g of fresh material, the determinations being carried out in three repetitions.

The total polyphenols were expressed in mg of gallic acid equivalents and the highest concentration was Ulmeni (1,003.06 mg/100 g sample) followed closely by Ardud (997.81 mg/100 g sample) and Zimbor (987.02 mg/100 g sample).

The total amount of flavonoids was determined using a colorimetric method based on AlC13, and the concentration was expressed in mg of quercitin per 100 g of the fresh sample. The highest concentrations of total flavonoids were obtained at Cetariu (46.88 mg/100 g sample), Ulmeni (34.49 mg/100 g sample) and Ardud (25.80 mg/100 g sample).

For the determination of antioxidant capacity, the ABTS + method was used, which was also a colorimetric method, and the expression was made in equivalents of μM Trolox, a Vitamin E analog, which has a very strong antioxidant capacity. The highest antioxidant capacity of the fruits was recorded in Cetariu (42.75 μM Trolox/g sample), followed closely by Ulmeni (42.30 μM Trolox/g sample) and Galații Bistriței (34.97 μM Trolox/g sample).

Table 1.6 The concentration of bioactive compounds from blueberry fruits

Sample no.	***The location of plantations***	***Total polyphenols* (mg AG/100 g probă proaspătă)**	***Total flavonoids* (mg QE/100 g probă proaspătă)**	***Antioxidant activity* (μM Trolox/g probă proaspătă)**
1	Șintereag	666.45	10.62	7.46
2	Galații Bistriței	438.84	15.06	34.97
3	Cetariu	683.77	46.88	42.75
4	Mărișel	481.00	20.87	13.21
5	Ulmeni	1003.06	34.49	42.30
6	Copalnic	585.50	12.64	18.84
7	Ardud	997.81	25.80	23.34
8	Zimbor	987.02	19.34	31.32

Many natural antioxidants, especially flavonoids, have a wide range of biological effects, including antibacterial, antiviral, anti-inflammatory, anti-allergic, anti-thrombotic and vasodilator actions. Antioxidant activity is a fundamental feature of major importance for human health, many of the body's basic functions, such as anti-mutagenic, anti-carcinogenic or anti-aging activity, originate from this property (Velioglu 1998).

Blueberries have been reported to have a pharmacological impact against ophthalmic disorders, bioactive compounds in the affine improving the bloodstream and the amount of oxygen in the eye. The content of substances with a high antioxidant capacity also helps to protect the bone and muscular system, has anti-diabetic properties; the consumption of cranberry fruits can be attributed to lower blood pressure, decreased blood cholesterol and therefore lowering the risk of cardiovascular accident (Skrankova et al. 2015).

Regarding the highest total polyphenols content, the parcels of Ulmeni, Ardud and Zimbor record significant values. From the point of view of the benefits to human health, due to the high content of total polyphenols, the three parcels present those characteristics of the soils that determine the conditions for obtaining the highest quality fruits in terms of quality. The blueberry crops from Galații Bistriței and Mărișel obtained the lowest concentrations of total polyphenols, suggesting that the land does not have the characteristics necessary to obtain high concentrations of the bioactive compounds.

The blueberry plantations from Cetariu, Ulmeni and Ardud have the highest concentrations of total flavonoid content, and this is also explained by soil characteristics in the area. The lowest total flavonoid concentrations were recorded in Șintereag and Copalnic.

In order to identify the best crops for blueberry culture, the determination of areas where the highest concentrations of bioactive compounds in the fruit have been obtained is not sufficient as these concentrations increase as a response to environmental stress indicating that the plant is struggling with agents (water excess/deficiency, low soil nutrient concentrations, unfavorable climatic factors–excess sunlight, etc.).

Therefore, in order to determine the areas where the blueberry fruits present the best qualitative properties but also to identify the most favorable land for this crop, other studies have been carried out, regarding the suitability of the land for the blueberry culture and the determination of the influence of the different soil properties on the concentration of the compounds bioactive from blueberries.

Environmental factors have a major effect on the content of polyphenols. These factors can be pedo-climatic (pH, amount of soil nutrients, soil type, sun exposure, rainfall) or agronomic (greenhouse or field culture, biological culture, hydroponic culture, the yield of fruit on the shrub, etc.).

The results of own research, as well as the comparative studies from the literature, show that the properties of the different studied soil types impart different values regarding the content of polyphenols and flavonoids but also the antioxidant capacity of the fruits (Huang et al. 2012, Denardin et al. 2015). The results obtained also take into account the methods used for determinations as well as the climatic and pedological factors specific to the area.

Regarding the total flavonoid content and the antioxidant capacity of the fruit, it was noted that the lower level of nitrogen supply leads to an increase in the concentration of the bioactive compounds in the fruit. Such situations have been encountered in typical luvosol (Cetariu–very low nitrogen level and the highest content of total flavonoids–46.88 mg QE/100 g fresh sample), eutricambosol (Ulmeni–low nitrogen content and content slightly lower total flavonoids than Cetariu) and preluvosol (Ardud–mean nitrogen level and mean value of total flavonoid content–25.80 mg QE/100 g fresh sample).

The same correlation is also observed with the antioxidant capacity of the fruit, lower nitrogen content in the soil leading to the increase of this concentration, the occurrence of samples taken from the following types of soils: typical luvosol (Cetariu–very low nitrogen but the highest antioxidant capacity of fruits–42.75 µM Trolox/g fresh sample), eutricambosol (Ulmeni–low nitrogen level and the average value of antioxidant capacity of fruits) and aluviosol (Galații Bistriței–medium nitrogen level and lower value of antioxidant capacity of fruit–34.97 µM Trolox/g fresh sample). Analyzing the antioxidant capacity of blueberry fruits, the research carried out shows that there are no significant differences between the eutricambosol soil type with 42.30 µM Trolox/g ABTS fresh sample and preluvosol soil type with 42.75 µM Trolox/g (Ulmeni: 1.44 g/kg and Cetariu, 1.08 g/kg) as well as in the samples collected from the billon (Ulmeni: 1.34 g/kg and Cetariu 0.90 g/kg).

Different levels of nitrogen supply have an inversely influential effect on total fruit polyphenol concentrations, a lower level of nitrogen leading to higher concentrations of such bioactive compounds, and in this study, this can be noticed for euricambosol (Ulmeni–the lowest level of nitrogen and the highest concentration of total polyphenols 1,003.06 mg AG/100 g of fresh sample), preluvosol (Ardud–mean nitrogen and 997.81 mg AG/100 g fresh sample) and preluvosol Zimbor–middle nitrogen and 987.02 mg AG/100 g fresh sample).

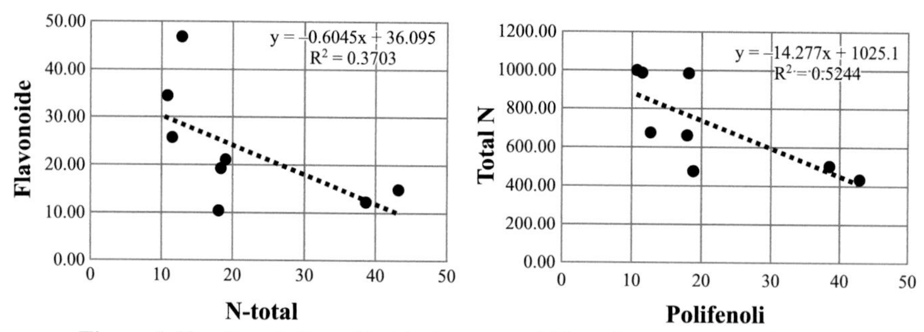

Figure 1.10 Correlation of total nitrogen and bioactive compounds in fruits (Pearson Correlation).

The analyzes performed in this study showed that there is also an influence between total pH and flavonoids, namely the antioxidant capacity of the fruits, a more alkaline pH increasing the concentrations of these fruit compounds. Influences have been observed for the same types of soils for which the total nitrogen level also has the same inverse relationship.

Based on the Pearson Correlation (Pearson Correlation) correlation, Figure 1.10 reproduces the correlations between total soil nitrogen level and total polyphenol and total flavonoid concentrations for all eight blueberry plantations studied. The application of this statistical test supports the results obtained in this scientific approach as well as the results obtained by other authors, which revealed the same types of correlations for these elements (Velioglu 1998).

Even though it is very important to determine the properties of soil for various cultivations and namely for the studied species, this study revealed that is also very important to know how the properties of soil can influence the qualitative properties of specie and the crop yield. The research carried out in the Northwest Region of Development from Romania highlights the importance of a combined study, the results obtained were very important for not only researchers but for farmers as well.

REFERENCES

Arnao, M.B., Cano, A., Alcolea, J.F. and Acosta, M. 2001. Estimation of free radical-quenching activity of leaf pigment extracts. Phytochemical Analysis. 12(2): 138–143. doi: 10.1002/pca.571

Bilaşco, Ş.T., Horvath, C., Cocean, P., Sorocovschi, V. and Oncu, M. 2009. Implementation of the usle model using gis techniques. Case study the somesean plateau. Carpathian Journal of Earth and Environmental Sciences. 4: 123–132.

Borlan, Z. and Hera, C. 1973. Metode de Apreciere a Stării de Fertilitate a Solului în Vederea Folosirii Rationale a Îngrăşămintelor, Ed. Ceres, Bucureşti, p. 487.

Borlan, Z., Răuţă, C., (Red. Coord.), 1981, Metodologie de Analiză Agrochimică a Solurilor în Vederea Stabilirii Necesarului de Amendamente şi Îngrăşăminte, vol 2. Seria Metode, raporte şi îndrumări, ICPA, 3: 489.

Boţ, A. 2017. Assessment of land favorability from Northwest Region of Development for blueberry crop (*Vaccinium corymbosum*), in the context of sustainable development of the territory, PhD thesis.

Bunea, A., Rugină, O., Pintea, A.M., Scoanţa, Z., Bunea, C.I. and Socaciu, C. 2011. Comparative polyphenolic content and antioxidant activities of some wild and cultivated blueberries from romania. Not Bot Horti Agrobo. 39(2): 70–76.

Cardeñosa, V., Girones-Vilaplana, A., Muriel, J.L., Moreno, D.A. and Moreno-Rojas, J.M. 2016. Influence of genotype, cultivation system and irigation regime on antioxidant capacity and selected phenolics of blueberries (*Vaccinium corymbosum* L.). Food Chemistry. 202: 276–283.

Clapa, D. 2006. Micropropagarea afinului cu tufă înaltă (*Vaccinium corymbosum*)- o soluţie rapidă pentru valorificarea solurilor acide, Agricultură–Ştiinţă şi Practică. nr. 1–2: 57–58.

Denardin, C.C., Hirsch, G.E., Rocha, R., Vizzotto, M., Moreira, J., Guma, F. and Emanuelli, T. 2015. Antioxidant capacity and bioactive compounds of four Brazilian native fruits. Journal of Food and Drug Analysis. 3(3): 387–398.

Diaconeas, Z., Leopold, L., Rugină, D., Ayvaz, H. and Socaciu, C. 2015. Antiproliferative and antioxidant properties of anthocyanin rich extracts from blueberry and blackcurrant juice. International Journal of Molecular Sciences. 16: 2352–2365.

Diaconeasa, Z., Ranga, F., Rugină, D., Cuibus, L., Socaciu, C. 2014. HPLC/PDA–ESI/MS Identification of phenolic acids, flavonol glycosides and antioxidant potential in blueberry, blackberry, raspberries and cranberries. Journal of Food and Nutrition Research. 2(11): 781–785.

Dragovic-Uzelac, V., Savic, Z., Brala, A., Levaj, B., Bursac-Kovacevic, D. and Bicko, A. 2010. Evaluation of phenolic content and antioxidant capacity of blueberry cultivars (*Vaccinium corymbosum* L.) grown in the Northwest Croatia. Food Technology Biotechnology. 48(2): 214–221.

Filipov, F. 2005. Pedologie, Ed. Ion Ionescu de la Brad, Iaşi, p. 444.

Florea, N. and Munteanu, I. 2012. Sistemul Român de Taxonomie a Solurilor, Ed. ESTFALIA, Bucureşti, p. 207.

Florea, N., Bălăceanu, V., Răuţă, C. and Canarache, A. 1986. Metodologia elaborării studiilor pedologice, vol I, II, III, Institutul de Cercetări pentru Pedologie şi Agrochimie, Bucureşti, p. 292.

Forney, C., Kalt, W., Jordan, M., Vinqvist-Tymchu, M. and Fillmore, S. 2012. Blueberry and cranberry fruit composition during development. Journal of Berry Research. 2: 169–177. DOI:10.3233/JBR-2012-034.

Gibson, L., Vasantha Rupasinghe, H.P., Forney, C. and Eaton, L. 2013. Characterization of changes in polyphenols, antioxidant capacity and physico-chemical parameters during lowbush blueberry fruit ripening. Antioxidant. 2: 216–229. DOI:10.3390/antiox2040216.

Halder, J.C. 2013. Land suitability assessment for crop cultivation by using remote sensing and GIS. Journal of Geography and Geology. 5(3): 65–74. DOI:10.5539/jgg.v5n3p65.

Howard, L.R., Clark, J.R. and Brownmiller, C. 2003. Antioxidant capacity and phenolic content in blueberries as affected by genotype and growing season. Journal of the Science of the Food and Agriculture. 83: 1238–1247. DOI:10.1002/jsfa.1532; http://lib.dr.iastate.edu/rtd/6234

Huang, W.-Y., Hong-Cheng, Z., Wen-Xu, L. and Chun-Yang, L. 2012. Survey of antioxidant capacity and phenolic composition of blueberry, blackberry, and strawberry in Nanjing. Journal of Zhejiang University, Science B. 13(2): 94–102. DOI: 10.1631/jzus.B1100137.

Kazim, G., Sedat, S. and Hancock, J.F. 2015. Variation among highbush and rabbiteye cultivars of blueberry for fruit and phytochemical characteristics. Journal of Food Composition and Analysis. 38: 69–79.

Kim, D.-O., Chun, O.K., Kim, Y.J., Moon, H.-Y. and Lee, C.Y. 2003. Quantification of polyphenolics and their antioxidant capacity in fresh plums. Journal of Agricultural and Food Chemistry. 51: 6509–6515.

Koca, I. and Karadeniz, B. 2009. Antioxidant properties of blackberry and blueberry fruits grown in the Black Sea Region of Turkey. Scientia Horticulturae. 12: 447–450.

Krasovskaya, V. 2012, Antioxidant Properties of Berries: Review of Human Studies and their Relevance in the Context of the European Food Safety Authority. Hogeschool van Amsterdam. https://pdfs.semanticscholar.org/3fac/75e90d1da26b2dc900bf147da543a98ad0a1.pdf

Łata, B., Trampczynska, A. and MIKE, A. 2005. Effect of cultivar and harvest date on thiols, ascorbate and phenolic compounds content in blueberry, Warsaw Agricultural University. Acta Scientarum Polonorum Hortorum Cultus. 4(1): 163–171.

Mehvesh M. and Wani, S.M. 2013. Polyphenols and human health—A Review. International Journal of Pharma and Bio Sciences. 4(2): 338–360.

Mladin, P. and Ancu, I. 2014. Tehnologia afinului cu tufa înaltă, Institutul de Cercetare-Dezvoltare pentru Pomicultură Piteşti. https://www.icdp.ro/laboratoare/arbusti/images/3.4.Tehnologie%20afin/Tehnologia%20de%20cultura%20a%20afinului%20cu%20tufa%20inalta.pdf

Moldovan, N., Bilaşco, Şt., Roşca, S., Păcurar, I., Boţ, A.I., NegruŞier, C., Sestraş, P., Bondrea, M. and NAŞ, S. 2016. Identification of land suitability for agricultural use by applying morphometric and risk parameters based on GIS spatial analysis. Not Bot Horti Agrobo. 44(1): 302–312. DOI:10.15835/nbha44110289

Păcurar, I. and Buta, M. 2010. Pedologie şi Bonitarea Terenurilor Agricole, Lucrări Practice, Ed. AcademicPres, Cluj-Napoca, p. 200.

Păcurar, I. 2000. Pedologie Generală şi Bonitarea Terenurilor Agricole, Ed. Academic Pres, Cluj-Napoca, p. 223.

Qi, Y., Baowu, W., Feng, C., Zhiliang, H., Xi, W. and Pengju, G.L. 2011. Comparison of anthocyanins and phenolics in organically and conventionally grown blueberries in selected cultivars. Food Chemistry. 125: 201–208.

Roşca, S. 2014. Application of soil loss scenarios using the romsem model depending on maximum land use pretability classes. A case study. Studia UBB Geographia. LIX 1: 101–116.

Roşca, S., Bilaşco, Ş, Petrea, D.T., Vescan, I. and Fodorean, I. 2015. Comparative assessment of landslide susceptibility. Case study: The Niraj river basin (Transylvania depression, Romania). Geomatics Natural Hazard and Risk. 7(3): 1043–1064. DOI:10.1080/19475705.2015.1030784

Routray, W. and Ors, V. 2011, Blueberries and their anthocyanins: Factors affecting biosynthesis and properties. Comprehensive Revision Food Science and Food Safety. 10: 303–320. DOI: 10.1111/j.1541-4337.2011.00164.x.

Singleton, V.L., Orthofer, R. and Lamuela-Raventós. 1999.Analysis of total phenols and other oxidation substrates and antioxidants by means of folin-ciocalteu reagent. Methods in Enzymology. 299: 152–178.

Skrovankova, S., Sumczynski, D., Mlcek, J., Jurikova, T. and Sochor, J. 2015. Bioactive compounds and antioxidant activity in different types of berries. International Journal of Moleculra Sciences. 16: 24673–24706. doi:10.3390/ijms161024673.

Stevenson, D. and Scalzo, J. 2012. Anthocyanin composition and content of blueberries from around the world. Journal of Berry Research. 2: 179–189. DOI:10.3233/JBR-2012-038.

Su, Min-Sheng and Chien, Po-Jung. 2007. Antioxidant activity, anthocyanins, and phenolics of rabbiteye blueberry (*Vaccinium ashei*) fluid products as affected by fermentation. Food Chemistry. 104(1): 182–187.

Velioglu, Y.S., Mazza, G., Gao, L. and Oomah, B.D. 1998. Antioxidant activity and total phenolics in selected fruits, vegetables, and grain products. Journal of Agricultural and Food Chemistry. 46: 4113–4117.

Zhishen, J., Mengcheng, T. and Jianming, W. 1999. The determination of flavonoid contents in mulberry and their scavenging effects on superoxide radicals. Food Chemistry. 64: 555–559.

***Monografia geografică a României, 1960, Vol I–Harta solurilor României la scara 1:1500000;

***Sistemul American de Clasificare a Solurilor (USDA- Soil Taxonomy, 1999).

***Baza Mondială de Referinţă pentru Resursele de Sol (WRB-SR 1988);

*** ICPA (1987), Metodologia de bonitare a terenurilor, Vol. I şi II;

***unfccc.int/resource/docs/2015/cop21/eng/10a01.pdf

Chapter 2

Nitrogen in the Peats of Western Siberia

Lydia Ivanovna Inisheva* and Tatyana Vasiljevna Dementieva
Tomsk State Pedagogical University, Russia.

INTRODUCTION

The high content of nitrogen compounds determines the agricultural value of peat soils and peats (Cambell 1978). According to Tyurin's calculations (Tyurin 1965), nitrogen reserves in peat soils are close to nitrogen reserves in black chernozem, and according to Efimov and Tsarenko (Efimov and Tsarenko 1992) calculations, they considerably exceed them. However, more than 90% of the peat nitrogen is in the form of complex organic compounds and is available to plants only after their mineralization. The share of the most stable components of organic matter (OM) of peat—humic acids and lignin—accounts for 40–50% and 23–34% of total nitrogen, respectively. According to the data of Shirokikh (1981), the nitrogen of humic acids is 30–37% of total nitrogen (it is 23–26% in humic acids and 11–14% in fulvic acids), the nitrogen of non-hydrolyzable residue accounts from 38 to 45%. The high potential but low actual fertility of peat soils and peats are explained by these properties (Tishkovich 1978). However, this does not exclude the use of peat as an agrochemical raw material in the technological processes of obtaining peat production from it.

In the literature data (Lishtvan and Korol 1975), the total nitrogen content in peat is 0.7–4.1% of its organic mass, and in Lupinovich and Golub's opinion (1952), nitrogen content cannot exceed 4.5%, which the authors explain by the limit of the nitrogen content in humic acids.

*For Correspondence: inisheva@mail.ru

Some idea of the strength of fixing nitrogen compounds in peat can be obtained using the method of fractioning it by means of solvents with an increasing extracting effect. According to Tyurin (1965), the method of fractioning based on the hydrolysis of OM by acids of different concentrations is the most suitable one for determining the amount of nitrogen of OM capable of decomposition since it is close to natural processes occurring in the soil with the participation of hydrolytic enzymes.

Unfortunately, the authors use different, often original methods of the fractionation of nitrogen compounds (in terms of quality, the concentration of leaches, and extraction regimes) in the research study of the nitrogen stock of peat and peat soils. This makes it difficult to systematize and compare the results obtained.

The method of fractioning developed by Bremner is the most commonly used as well as the variant of Shkonde and Koroleva (Methods for the Determination of Nitrogen in the Soil, 1975). In the work of Efimov and Tsarenko (1992), the results of the analysis of the fractional composition of nitrogen of one peat sample which was obtained by these two methods were presented. Based on these data, it can be concluded that the differences of the methods are reduced not only to different names of the released fractions (amino acid fraction, hexosamine and ammonium fractions, hydrolyzed by the first method and mineral, easily hydrolyzable, hardly hydrolyzable by the second method) but also by the amount of nitrogen of the non-hydrolizable residue (almost 2 times–respectively 17 and 29% of the total nitrogen). In addition to this, the Bremner method is not very accurate. While the method of Shkonde-Koroleva, being more accurate, provides information on the mobility of all nitrogen compounds of the peat soil (Methods for the Determination of Nitrogen in the Soil, 1975). Thus, the latter method is more informative in studies related to the assessment of the nitrogen stock of peat soils and peats. Also, it allows estimating the reserves of nitrogen compounds in peats at different levels of the OM transformation.

In order to identify the features of the distribution of nitrogen compounds in peats, depending on their genesis, we determined the fractional composition of nitrogen in West Siberian peats using the Shkonde-Koroleva method.

MATERIALS AND METHODS

The peculiarities of the botanical composition of the peats of Western Siberia allowed us to identify a number of peats not found on the European territory of Russia such as sogra peat, larch peat, firry peat, sedge-grass, and pine-sedge peats, etc. In Western Siberia, the botanical classification includes 186 species of peats (Classification of Vegetation Cover And Peat Species in the Central Part Western Siberia 1975) related to three types and 18 groups (Figure 2.1). Of these, the stratified peats are represented by 76 ones. For example, in the Tomsk region, ten types of peat are absolutely dominant: five species of raised bog peat (Sphagnum fuscum peat, Angustifolium peat, Sphagnum medium peat, Sphagnum peat of bog hollows and complex high bog peat) and five types of lowland swamp peat (woody

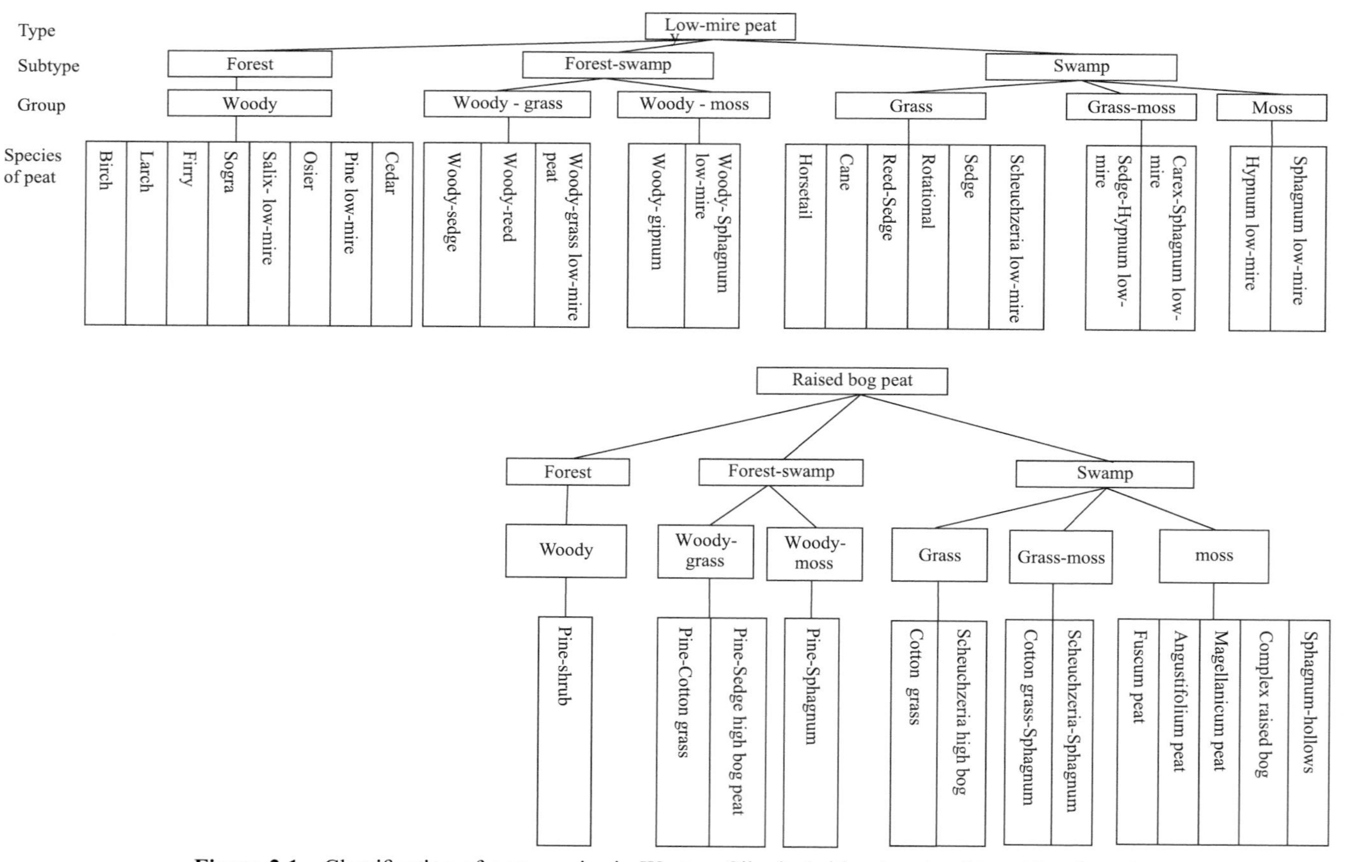

Figure 2.1 Classification of peat species in Western Siberia (without peats of transitional type).

peat, woody-grass peat, sedge peat, sedge-hypnum, and Hypnum lowland swamp peats) (Liss et al. 2001).

Within the limits of the taiga zone of Western Siberia, expeditionary research was conducted on representative peat deposits. In the central parts of the peat deposits, peat samples were taken in which chemical analysis was performed.

The studied samples belong to 12 types of peat (of raised bog and lowland swamp types), including all types of lowland swamp peats representative for Western Siberia (woody peat, woody-sedge, woody-grass, sedge peat, grass peat, sedge-Hypnum, and Hypnum low-mire peats) as well as the main types of raised bog peats (Fuscum, complex, Sphagnum peat of bog hollows, Carex-sphagnum, and Scheuchzeria). Each type of peat, according to its botanical composition, is represented by the representative except 6–19 samples taken from peat deposits of different swamps of Western Siberia.

The fractional composition of nitrogen was determined by the Shkonde-Koroleva method (Methods for the Determination of Nitrogen in the Soil, 1975). According to this method, peat nitrogen is divided into 4 groups:

1. Mineral nitrogen (nitrogen of nitrates, nitrites, and exchangeable ammonium);
2. Easily hydrolyzable nitrogen (amides, a part of amines, a part of non-exchangeable ammonium), recoverable when processing the soil with 0.5 n. sulfur acid and representing the distilled off a fraction of the nitrogen of this hydrolyzate;
3. Hardly hydrolyzable nitrogen (a part of amines, amides, non-exchangeable ammonium, a part of humins) which is a distilled off part of soil nitrogen extracted by 5 n. sulfur acid;
4. Non-hydrolysable nitrogen (a significant part of amines, humines, melanins, bitumens, non-exchangeable ammonium residue) which represents the amount of non-distilled nitrogen of the hydrolyzate 5 n. sulfur acid and non-hydrolysable nitrogen. The analyzes were carried out in the testing laboratory of the Tomsk State Pedagogical University (the accreditation certificate No.ROSS RU.0001.516054).

The results were statistically processed using the STATISTICA 6 application package.

RESULTS AND DISCUSSION

In the studied samples of raised bog peats, the total nitrogen content varies within 0.64–1.99% with an average content of 1.26% calculated on the absolutely dry mass (Table 2.1). Lowland swamp peat contains 0.43–2.27% with an average grade of 1.94%. Higher values of total nitrogen in lowland swamp peat types are determined among other things by the content of mineral nitrogen compounds accumulating in lowland swamps as well as by the peculiarities of eutrophic peat-forming plants. The lowland grass peat group is the most nitrogen-enriched (on average, 2.22%) followed by the wood peat group (with 1.91%) and the moss peat group (1.43%). Lowland swamp wood-grass and grass-moss peat groups occupy an intermediate position.

In raised bog peats, the total nitrogen content is much lower (Table 2.1). The moss peat group has the lowest average total nitrogen content (1.03%). The Fuskum peat in which the nitrogen content does not exceed 0.64% stands out in this group. The peat of grass-moss group contains on average 1.28% of total nitrogen. The grass peat group (Scheuchzeria) is characterized by a high content of total nitrogen with 1.99% on average with limits of 1.91–2.09%, which is explained by the increased content of humic acids in these peats. By the properties of OM, this approximates the peat of the grass group to lowland swamp peats.

Table 2.1 Fractional composition of nitrogen in peats of different botanical composition (Note: Cv is the coefficient of variation; % a.d.m.—percent of absolutely dry mass; the contents of the mineral, easily hydrolyzable, hardly hydrolyzable and non-hydrolyzable nitrogen given in percents of the total nitrogen)

Botanical composition	***Total, %* a.d.m.**		***Mineral***		***Easily hydrolyzable***		***Hardly hydrolyzable***		***Non-hydrolizable***	
	mean	Cv,%	mean	Cv,%	mean	Cv,%	mean	Cv,%	mean	Cv,%
Low mire type										
Group woody	1.91	16.5	1.61	215	9.05	21.8	8.21	48.1	80.89	6.5
Group woody-grass	1.99	27.2	1.10	69	8.84	33.0	6.18	77.3	83.54	7.2
Type woody-grass	1.99	25.5	1.31	59	9.13	29.2	6.09	89.7	83.21	8.4
Group grass	2.22	21.4	1.43	79	9.17	30.8	5.77	48.4	83.55	4.4
Type sedge	2.16	24.7	1.19	97	8.6	29.0	6.37	42.7	83.72	3.9
Type grass	2.27	19.0	1.64	67	9.65	31.8	5.27	53.7	83.40	4.9
Group grass-moss	1.71	44.3	1.97	78	10.27	61.7	6.53	58.1	81.22	10.3
Group moss (hypnum)	1.43	33.1	1.99	72	13.39	45.6	9.44	46.1	75.18	14.4
mean on type	1.94	29.0	1.56	116	9.83	42.0	6.91	57.0	81.57	9.0
Raised bog type										
Type scheuchzeria	1.99	4.0	1.60	22	7.42	17.1	5.85	15.9	85.00	2.2
Type cotton-Sphagnum	1.24	38.8	8.02	78	11.02	34.3	6.63	58.4	75.12	12.7
Group grass-moss	1.28	35.6	7.25	84	10.32	38.1	6.67	54.6	76.43	12.3
Group moss	1.03	42.2	4.82	35	7.58	46.5	6.72	36.5	80.42	5.0
Type fuscum	0.64	19.3	5.60	25	6.07	56.2	8.77	30.7	79.53	4.6
Type complex	1.11	40.7	4.64	43	9.27	49.9	6.43	32.9	79.59	6.2
mean on type	1.26	41.0	5.24	80	8.49	43.0	6.67	41.0	79.60	9.0

The share of each of the fractions in total nitrogen (% of N total) allows us to estimate the degree of stability of nitrogen compounds in peat to biochemical destruction and it is given in Table 2.1. The contents of fractions in mg/kg of absolutely dry matter (a.d.m.) of peat characterize the total nitrogen reserves of each fraction in peats and peat soils of different genesis, determining their fertility level. These values are given directly in the text of the article.

Thus, the mineral nitrogen compounds are characterized by an insignificant content from the total nitrogen: by 1–2% in lowland swamp peats and 1.6–8% in raised bog peats. Typically, the stocks of this fraction in peats do not exceed

100 mg/kg a.d.m. Moreover, the average values of the content of mineral nitrogen by groups of peat of different botanical compositions do not differ especially. Mineral nitrogen is represented mainly by ammonium and nitrate fractions. These compounds are characterized by significant seasonal variability and therefore cannot be considered as a parameter associated with the genesis of peat.

The content of easily hydrolyzable nitrogen is on average 9.8% of the total nitrogen for the lowland swamp peat and 8.5% for raised bog peat. Differences in the intervals of variation (0.5–28.8% and 0.7–18.0%, respectively) and the average values by the types of peat are insignificant, while peat groups of different botanical compositions have more pronounced differences in the content of easily hydrolyzable nitrogen. The total reserves of readily hydrolyzable nitrogen vary from 6 to 429 mg/kg, lowland swamp peat contains on average 185.1 mg/kg, and raised bog peat–only 105 mg/kg of a.d.m.

The lowest content of easily hydrolyzable nitrogen is in the fuskum peat of raised bog peat type and sedge peat in the lowland swamp peat type. The lowland swamp peat of moss group is characterized with 221 mg/kg of easily hydrolyzable nitrogen (Nhe) as well as the peat of grass and moss groups of raised bog and lowland swamp peat types are characterized by the highest content of easily hydrolyzable nitrogen; this defines their nitrogen as potentially more mobile. The easily hydrolyzable fraction of nitrogen is considered as the nearest reserve for plant nutrition (Zamyatina 1975, Efimov and Tsarenko 1992). However, there is an opinion (Pereverzev 1987) that in climatic conditions, when the mobilization of soil nitrogen is difficult because of the low activity of microflora, high content of easily hydrolyzable nitrogen does not indicate large reserves of nitrogen available to plants apparently due to the low rate of nitrogen conversion to the forms which are assimilable by plants and their insufficient accumulation. Nevertheless, this fraction of nitrogen in the transformation of the peat organic matter will be primarily involved in the transformation of nitrogen compounds.

The content of hardly hydrolyzable nitrogen varies from 5 to 9% of the total nitrogen, yielding to the content of easily hydrolyzable forms. There were no significant differences in the content of hardly hydrolyzable nitrogen by the types of peat. However, for the lowland swamp peat, a larger interval of variation of indicators is characteristic (Table 2.1). The raised bog groups of peats are characterized by close average values of the content of hardly hydrolyzable nitrogen, while in peats of lowland swamp type an increase in the proportion of hardly hydrolyzable nitrogen from the grass peat group to the woody and moss ones is observed. The peats of the lowland swamp woody group are characterized by the highest content of this form of nitrogen (391 mg/kg of a.d.m.).

The fraction of non-hydrolyzable nitrogen, reaching 50.6–91% of the total nitrogen in the studied peats, is predominant. A high percentage of the non-hydrolyzable nitrogen fraction is determined by the presence of compounds inaccessible to plants and microorganisms.

It should be noted that in groups of peats of different botanical composition, the content of non-hydrolyzable nitrogen does not change significantly. Lowland swamp peats contain non-hydrolyzable nitrogen in average values of 81.6%, and raised bog peats contain it up to 79.6%. In the lowland swamp peat type, the moss group stands out, where the average non-hydrolyzable nitrogen content is only 75% with a fluctuation range of 51–91%, while in the raised bog peat type, the grass group (Sheuchzeria raised bog peat) is with the highest non-hydrolyzable nitrogen content of 85%. The total reserves of this fraction vary within 92–2,499 mg/kg of a.d.m.

The lowest content of non-hydrolyzable nitrogen is characteristic for raised bog peats of moss and grass-moss groups (837–1,005 mg/kg on a.d.m.). The lowland swamp peat of the moss group is approaching the same level with an average of 1,094 mg/kg. Grass groups of both types have the highest content of the non-hydrolyzable nitrogen with an average of 1,689–1,897 mg/kg. A general pattern is revealed: moss peats contain less of the non-hydrolyzable nitrogen than grass ones, and raised bog peats contain less of it than lowland swamp peats.

Particularly noteworthy is the degree of enrichment of the peats OM with nitrogen, estimated by the value of C:N since it is known that it has a significant impact on the intensity of microbiological destruction of OM.

If we estimate the degree of enrichment of the peat OM with nitrogen by Efremova (1990), according to whom the C:N ratio for more than 14 for peat soils and peats can be taken as very low, then the lowest enrichment of peats with total nitrogen is noted for the raised bog moss peat group (56 on average); the highest– for lowland swamp peats of grass group (19). Intertype differences are even more pronounced: fuscum peat relates C:N as 77.6 on average, whereas low-mire sedge peat–8.7 (Table 2.2).

Table 2.2 Influence degree of peat decomposition on relation C:N in peats

Botanical composition	*Number of observation*	*Mean*	*Minimum*	*Maximum*	C_V
		Low mire type			
Group woody	18	23.8	15.9	30.6	0.195
Group woody-grass	19	20.3	14.1	32.6	0.233
Type woody-grass	14	20.9	14.1	32.6	0.234
Group grass	35	19.3	14.3	32.7	0.197
Type grass	19	19.9	14.8	32.7	0.226
Type sedge	16	18.7	14.3	23.2	0.149
Group grass-moss	14	22.1	11.7	31.4	0.275
Group moss	14	26.6	20.9	38.7	0.180
		Raised bog type			
Type scheuchzeria	4	26.8	26.4	27.5	0.019
Type cotton-sphagnum	12	48.1	21.5	92.7	0.406
Group grass-moss	14	46.0	21.5	92.7	0.409
Group moss	19	56.3	27.5	99.8	0.381
Type fuscum	6	77.6	54.9	99.8	0.207
Type complex	7	51.7	28.3	77.7	0.310

Differences in C:N values are closely related to the degree of peat decomposition since carbon content decreases faster than nitrogen during plant residue transformation and peat formation.

We analyzed the average values of the C:N ratio of peats grouped by the peat types and the degree of its decomposition (Table 2.3).

Table 2.3 Impact of degree of decomposition on the ratio C:N in peats

Interval of peat decomposition degree %	*Number of observation*	*Mean*	*Minimum*	*Maximum*	*Dispersion*	*Cv*
			Low mire type			
Less 20	17	21.3	14.1	29.6	4.315	0.202
20–25	21	23.2	14.4	38.7	5.631	0.242
25–30	26	22.4	15.4	30.3	4.694	0.21
30–45	32	20.5	11.7	32.7	5.500	0.268
45–60	4	21.0	15.5	31.4	7.404	0.353
			Rraised bog type			
Less 10	8	69.4	28.3	99.8	28.107	0.405
10–15	13	51.2	21.5	72.8	15.453	0.302
15–25	10	40.1	26.4	57.8	13.048	0.325
25–36	8	31.9	26.5	48.6	7.352	0.23

Based on the obtained results, it follows that the above assumption about the interdependence of the degree of decomposition and the C:N ratio is valid only for raised bog type of peats; lowland swamp peats in the intervals of variation of the degree of decomposition from 12 to 56% are characterized by the same C:N ratio of 21–23, while the C:N ratio in raised bog peats decreases from 69 to 32 with increasing degree of decomposition from 0 to 35%. This C:N ratio in raised bog peats and its dependence on the degree of decomposition ascertains their increased capacity for microbiological decomposition and accordingly low biochemical stability.

CONCLUSIONS

1. The raised bog peats are less enriched with nitrogen than lowland swamp ones. The moss group of raised bog peats is characterized by the lowest nitrogen content, including easily hydrolyzable fractions. At the same time, the C:N ratio in these peats and its dependence on the degree of decomposition indicate an increased ability of raised bog peats to microbiological decomposition and chemical hydrolysis.
2. Among the raised bog peat types, Sheuchzeria peat is distinguished, which is sufficiently enriched in nitrogen. In terms of the fractional composition of nitrogen, the Scheuchzeria peat is approaching lowland swamp peat.
3. The groups of lowland swamp peats are characterized by an enlarged nitrogen content with an increase in the proportion of non-hydrolyzable fraction and a decrease in the share of mineral nitrogen. Among the groups of peat of the lowland swamp type, a moss group is distinguished with a low content of total nitrogen and its non-hydrolyzable fraction.

4. The grass-moss groups of both types of peats occupy an intermediate position between the grass and moss groups, except for the content of easily hydrolyzed nitrogen, which takes the highest values that characterize the lower resistance of the nitrogen compounds of these groups to transformation.
5. The results of the analysis of the fractional composition of nitrogen confirm the leading role of the botanical composition of peats in the distribution of nitrogenous compounds into fractions as Pereverzev (1987) and Bambalov and Yankovskaya (1994) have previously indicated.
6. Different peats are characterized by a wide range in the distribution of nitrogen compounds in fractions and consequently at the basis of their choice as natural agrochemical raw materials in agricultural production, the leading role belongs to the botanical composition and the degree of decomposition.

REFERENCES

Bambalov, N.N. and Yankovskaya, N.S. 1994. Fractional composition of the nitrogen pool of organic fertilizers and peat-forming plants. Agrochemistry. 8: 55–61 (in Russian).

Cambell, C.W. 1978. Soil organic carbon, nitrogen and fertility. *In*: Schnitzer, M. and Khan, S.U. (eds), Soil Organic Matter. Elsevier Publishing Company, New York, pp. 173–271.

Classification of vegetation cover and peat types in the central part of Western Siberia. 1975. M. Mingeo SSSR. Trest "Geoltorfrazvedka", p. 150 (in Russian).

Efimov, V.I. and Tsarenko, V.P. 1992. Organic matter and nitrogen from peat soils. Soil Science. 10: 40–48 (in Russian).

Efremova, T. 1990. Humus and structure formation in forest peat soils of Western Siberia. Author. diss. Dr. Biol. Sciences, Novosibirsk, p. 48 (in Russian).

Lishtvan, I.I., Bambalov, N.I. and Tishovich, A.V. et al. Their substances and their practical use. Chemisty of Hard Fuel. 6: 14–20.

Lishtvan, I.I. and Korol, N.T. 1975. The basic properties of peat and methods of their determination. Minsk: Science and Technology, pp. 320 (in Russian).

Liss, O.L., Abramova, L.I., Avetov, N.A., Berezina, N.V., Inisheva, L.I., Kurnishkova, T.V., Sluka, Z.A., Tolpysheva, T.Yu. and Shvedchikova, N.K. 2001. Marsh Systems, Their Conservation Value, p. 584 (in Russian).

Lupinovich, I.S. and Golub, T.F. 1952. Peat-bog Soils and their Fertility. Minsk. Science and Technology, p. 268 (in Russian).

Methods for the determination of nitrogen in the soil. 1975. Agrochemical methods of soil research. M. Science, pp. 94–95 (in Russian).

Pereverzev, V.N. 1987. Biochemistry of humus and nitrogen in soils of the *Kolskogo Peninsula* L. Science. p. 303 (in Russian).

Shirokikh, P.S. 1981. Organic matter and nitrogen compounds in lowland peat soils with different botanical composition. Siberian Journal of Agricultural Sciences. 1: 16–20 (in Russian).

Smólczyński, S. 2009: Physical properties of upper silted organic soils in various landscapes of northeastern Poland. *In*: Bieniek, B. (ed.), Soil of Chosen Landscapes. Monograph. Department of Land Reclamation and Environmental Management, University of Warmia and Mazury in Olsztyn, pp. 51–87.

Szajdak, L.W., Inisheva, L.I., Styla, K., Gaca, W. and Meysner T. 2012. Tagan Peatland. Necessity of Peatlands Protection. Institute for Agricultural and Forestry, Polish Academy of Science, Poznan, pp. 367–377.

Tishkovich, A.V. 1978. The Properties of Peat and the Effectiveness of its Use for Fertilizer. Minsk. Science and Technology, p. 152 (in Russian).

Tyurin, I.I. 1965. Soil Organic Matter and Its Role in Fertility. Izd. Nauka, p. 320 (in Russian).

Chapter 3

Soil Tillage and Denitrification: Source from Depollution the Soil, Subsoil and Aquifer Water—Description, to Conduct and Control

Teodor Rusu* and Paula Ioana Moraru

University of Agricultural Sciences and Veterinary Medicine Cluj-Napoca, Romania.

INTRODUCTION

In any agricultural system, it is important to establish the meaning and the rate of modifications that take place in the soil under the influence of different technological systems and cultural practices. Durable agriculture approaches in a new ecological-energetic concept the processes and phenomena which represent the most active interface between living and non-living on our planet—the upper part of the Earth's surface, the soil, the subsoil and the aquifer water. In this context, it is necessary to find solutions and solve by research the nitrate pollution, stop the decline of the organic matter, reduce erosion and soil degradation.

Due to weather difficulties and a lot of time needed for the research on the synthesis and decomposition of organic substances, mineralization, accumulation and loss of nitrogen from the quantity and quality point of view, the experimental data at an international level is fragmental and does not cover the entire diversity of situations met in the agricultural practice. Based on the main facts of these problems, the current paper makes an analysis of these.

*For Correspondence: E-mail: trusu@usamvcluj.ro.

This paper summarizes the results of the researches performed and quoted in the literature, supplemented with the results from the researches conducted at the University of Agricultural Sciences and Veterinary Medicine Cluj-Napoca, Romania, regarding the possibility of controlling and directing the processes of nitrification-denitrification from the soil through factors with direct influence like the existing microflora, the level anoxia, availability in oxidized nitrogen and carbon substrate or factors with indirect influence such as soil tillage system, soil humidity, application of fertilizers and organic matter.

THE MAIN ECOLOGICAL FUNCTIONS OF THE SOIL

The soil is the main important component of the Earth's biosphere, a role resulted from the complex ecological functions it has inside it as well as by the interaction of biotic and abiotic components that determine its fertility and form the bioproductive systems.

The soil has at least 10 important functions inside the biosphere (Rusu 2005):

1. The energetic function, ensuring the storage of an enormous energy potential produced through photosynthesis and accumulated in the organic matter from the soil.
2. The hydrological function through which the accumulation and circulation of water preserve between the atmosphere, soil and land waters are harmonized.
3. The biogeochemical function through which the retention, slow availability and circulation of elements essential to life are ensured.
4. The gas-atmosphere function through which the quantity and quality proportions of sequestration, aeration and release of gas between the soil and the atmosphere are fixed.
5. The production of biomass, ensuring food, feed, renewable energy and raw materials.
6. The function of filtration, buffer and transformation among the atmosphere, hydrosphere and biosphere, protecting the environment, including human beings, especially against the contamination of aquifer water and the food chain.
7. The soil is a biological habitat and gene tank containing more species in number and quantity than all the other biological environments together.
8. Using the soil as a base for technical, industrial and social-economical structures and their development (industrial, house constructions, communication paths, sports spaces, parks and waste ramps).
9. Using the soil as a source of raw materials (clay, sand, gravel, the ore in general, etc.) but also as a source of energy and water.
10. The soils are important geogenic and cultural artifacts, forming an essential part of the landscape where we live.

The soil is a dynamic entity, and dynamism means life. Being an open energetic system, extremely complex, polyphasic and polyfunctional, with its structure and organization, but functionally related to the environment through permanent flows of matter and energy; the modification of soil conditions causes the modification of its organization and functioning going through several steps—primary, development, balance and evolution. The soil dynamics or the complex modifications taking place in the soil are emphasized by many processes that continue to change until a certain relatively stable balance is reached. The evolution of the soil to the stationary state corresponding to the classical climax state (of balance with the environment) is applicable at a time scale of hundreds to thousands of years. Looking from a geological time scale, the soils, including the ground shells, are in continuous evolution, parallel with the evolution of environmental conditions, in permanent dynamics, submitted to geological variations (including climate).

Understanding the functions and management of the soil, including food production, storage, filtration and transformation of minerals, water, organic matter, gas, etc., as well as supplying raw materials are at the basis of human activity, the past, the present and especially the future of the Earth. The soil, "mirror of the landscape", reflects the entire evolution of the landscape, not only its current state. Each soil is a "memory block" which records past and present interactions between different geospheres, reflecting their history (Florea 2003).

Most of the time, man analyzes the energetic, biogeochemical, hydrological and gas-atmosphere functions from the point of view of reaching the bioproductive function without regarding the soil as a harmonious system with multiple ecological functions and without assessing in real-time its function of protecting the environment, respectively, the capacity of filtration, buffer and transformation among the atmosphere, hydrosphere and biosphere. This perspective leads to an accelerated evolution of the soil and especially to the appearance of degradations.

The energy accumulated in the soil is concentrated both in the mineral part of the soil as very stable energy with limited implications in the circulation of substances, as well as in plants, through the photosynthesis process and respectively the transformation into humus. By forming humus in the soil, great (secular) energy reserves are concentrated, which annually accumulate, creating a huge energetic potential of the soil. If we refer to the whole arable territory of Romania, the humus quantity represents almost a billion tons per 0–20 cm depth and two billion tons, per 0–50 cm depth. On average, agricultural lands in Romania have average humus reserves, the value of their weighted average is 136 ton/ha. One ton of humus equals, from an energetic point of view, 0.6 tons of petrol, 666.7 tons of gas and 1 ton of coal (Berca 2011).

Humus has also a great ecological importance by reducing the pollution of the environment with xenobiotic substances. Humus is a tank of nutritious elements for plants and microorganisms, which it puts at their disposal following the biochemical processes of mineralization, ion exchange and solubilization. The organic matter is the energetic substrate for life in the soil. Losing any part of humus means for the soil a corresponding reduction of its energetic function. Among the consequences of the negative results of humus in the soil, one can mention: structure degradation, intensification of compaction, reduction of permeability, increase of the need to apply mineral and especially organic fertilizers, the appearance of deficiency of certain microelements, decrease of activity of microorganisms, increase of the soil resistance to agricultural works, decrease of the buffer capacity, increase of the phytotoxic effect of certain heavy metals and erosion on slopes, decrease of irrigation efficiency, etc. Humus has also a bioproductive importance. Between the humus content and agricultural productions there is a positive correlation; the better the humus reserve, the higher the agricultural productions.

The function of filtration, buffer and transformation among the atmosphere, hydrosphere and biosphere is a very important one for the protection of the environment, especially as it is complex and depends on the stability, quality and nature of the macro and microporous system of the soil. The porous system of the soil is the one that controls the processes of transport of solutions to the plant through the radicular mass to groundwater or to surface waters through which absorbing toxic chemical components makes this environment act as a buffer and filtration system.

The agricultural technological systems through all their components (fertilization, irrigation, soil tillage, rotation, etc.) must ensure the improvement, protection and conservation of the porous system to maintain the quality state of the soil. The flora and fauna from the soil are the main components responsible for decomposing different organic and vegetal materials with transforming organic substances and toxic compounds.

THE BALANCE OF SOIL NUTRITION: MINERALIZATION AND HUMIFICATION

Fertilization is the act of applying fertilizers in order to increase soil fertility and ensure the substrate needed to obtain the vegetal production based on the principle of returning nutritious elements. The principle of returning nutritious elements extracted from the soil together with the crop refers to the fact that crop plants consume, during their vegetation period, the nutritious ions needed for their life and ultimately for the crop formation.

The principle of returning to the soil the nutritious elements which it was deprived of targets to maintain the fertility state at a high level (compared to the type and technology), capable of permanently ensuring vegetal productions. In order for a crop to produce at a quantity and quality level corresponding to its potential, under favorable environmental conditions, it must have, during the entire vegetation period, a series of mineral nutrients (nitrogen, phosphorus, potassium, calcium, magnesium, sulfur, iron, manganese, copper, zinc, boron, molybdenum and chlorine) in adequate quantities and proportions (Table 3.1 according to ICPA 2005).

The cycle of nutritious elements in the soil, the biochemical processes and the evolution of the soil fertility are influenced decisively by the microorganisms in the soil. In the soil, there is a great number of microorganisms but also macroorganisms as well as roots and other underground organs. In this community of living beings, there are very different relations: they live together, they succeed each other (metabiosis), they support each other (symbiosis) or some derive nutrients at the other's expense, etc. The processes that appear following the activity of

Table 3.1 Average consumptions (exports) of nutritious elements from the soil for the crop formation

Crop	***Main: Secondary production report***	***Exported nutritious elements, kg/ton of main crop and adequate quantity of secondary production***		
		N	P_2O_5	K_2O
Autumn wheat	Grains:Straws 1:1.3	26.5	13.7	16.4
Barley and two-row barley	Grains:Straws 1:1	23.0	10.8	22.3
Rye	Grains:Straws 1:1.5	27.5	9.4	26.8
Oat	Grains:Straws 1:1.5	28.5	11.0	31.2
Grain maize	Grains:Stems 1:1.6	27.5	12.5	16.5
Maize for silo	Whole plants with cobs	6.5	3.0	5.5
Sugar beet	Roots:Leaves and necks 1:1	4.9	2.0	6.0
Beet	Roots:Leaves 1:0.5	3.8	1.7	7.9
Potato	Tubers:Stalks 1:0.5	5.2	2.7	7.5
Sunflower	Seeds:Stems 1:3	36.5	17.5	50.0
Canola for oil	Seeds:Stems 1:3	51.5	36.0	44.0
Flax for seeds	Seeds:Stems 1:3	59.0	17.3	72.0
Beans	Grains:Stalks 1:1.5	59.5*	13.4	25.0
Peas	Grains:Stalks 1:1.5	61.0*	16.6	28.0
Soy	Grains:Stalks 1:1.5	70.0*	22.5	34.0
Flax tow	Stems	11.0	7.0	13.0
Hemp	Stems	10.0	8.5	17.5
Alfalafa	Green mass at the beginning of the blossom	8.0*	1.6	6.5
Red trefoil	Green mass at the beginning of the blossom	6.5*	1.5	5.5
Grass of natural meadows	–	6.5	1.4	4.5
Orchard grass	Green mass	6.0	1.7	8.3
Meslin (oat + vetch)	Green mass	6.5*	2.4	5.5
Maize	Green mass	3.0	1.7	4.5
Alfalfa hay	Beginning of the blossom	32.0*	6.4	22.0
Red trefoil hay	Beginning of the blossom	26.0*	6.0	21.0
Natural meadow hay	–	24.0	5.6	18.0
Hay of cultivated perennial grasses	–	23.0	6.5	28.0
Meslin hay (oat + vetch)	–	25.0*	8.0	20.0
Alfalfa hay mixed with ryegrass	–	26.0*	6.0	20.0
Apples	Fruit	1.6	0.5	2.0
Wine grapes	(+ Secondary production)	6.5	1.6	5.5
Tomatoes	Fruit	2.9	1.0	4.5
Autumn cabbage	Head of cabbage	3.5	1.2	4.0

*mostly coming from the symbiosis with the microorganisms which fix nitrogen

the organisms from the soil have a great influence on its fertility and therefore on vegetal production. Thus, from the multitude of biological and biochemical processes, especially dynamic, the soil and agrotechnical point of view, the following are essential:

- The process of humus formation (humification);
- The mineralization of the organic matter and release of elements used in the plant nutrition (ammonification, nitrification and denitrification);
- The enzyme activity of the soil;
- The relations between the plant roots and the microorganisms from the soil (fixation of atmosphere nitrogen);
- The relations between the microorganisms from the soil (commensalism, protocooperation, symbiosis, competition, amensalism, parasitism and predatorism);
- The relations between the roots of different plants (favorable or antagonist, allelopathy).

Humification (synthesis, humus accumulation) is the process according to which existing organic debris introduced in the soil are decomposed by microorganisms through slow biochemical oxidation into substances with reduced molecular mass, which are subsequently transformed through condensation and polymerization into substances with high molecular weight, rich in nitrogen and carbon that make the humus.

Mineralization (decomposing) of non-humified organic, intermediate substances or including a part of the humus takes place by the release of elements and compounds easily accessible to plants. Humus is an essential source of nitrogen to plants. Annually, by microbiological activity, an average of 0.5–1.0% that is 50–100 kg/ha N is mineralized. In the arable layer of the soils from the temperate area, the total content of organic N ranges on an average between 1–12 t/ha.

In the agricultural ecosystems, one must keep a balance between these two contrary processes and ensure energy to the soil by supplying it with organic and mineral material (the organic matter represents the energetic substrate, while the mineral fertilizers stimulate the growth of the radicular system and thus of the quantity of vegetal debris from the soil).

In contemporary agriculture, the report between the humification and mineralization of the organic matter depends on a series of factors among which the cultivated plant, the technology used, the quantity of vegetal debris left, the intensity of the soil tillage, the fertilizers used, the type of soil, the climate conditions, etc. The maintenance of humus quantity needs almost 8–10 t/ha of vegetal debris annually, ensured only under crops of perennial grasses. Under the conditions from Romania, both straw cereal and weeding plants leave in the soil annually 1–4 t/ha of vegetal debris. Consequently, the positive quantity of humus can be ensured only by supplying it with organic fertilizers.

The mineralization of organic substances with nitrogen is influenced by the soil, air, humidity, the C/N report of the organic substrate and its quantity, the existing microorganisms, etc. This is a very complex process, but schematically it can be presented in two steps (Blaga et al. 1996):

1. The ammonification or the formation of ammonia (NH_3) and other products (CO_2, H_2O, H_2S, etc.) by slow microbiological decomposition of the molecules in less complex groups.
2. The nitrification or the process according to which ammoniacal nitrogen is converted into nitric nitrogen. The nitrates are the main source of nitrogen for plants.

The transformation of ammonia into nitric nitrogen results schematically from the two following oxidation reactions:

a) The ammonia transformed into nitrous nitrogen: $2NH_3 + 3O_2 = 2HNO_2 + 2H_2O$ + energy (66 Kcal)
b) The nitrous nitrogen is passed into nitric nitrogen: $2HNO_2 + O_2 = 2HNO_3$ + energy (17 Kcal)

The nitrous and nitric acids combine quickly with different elements forming salts (Ca nitrate, K, Mg, Na, etc.). The nitrification bacteria are strictly autotrophic and aerobe; the process being a typical chemosynthesis one; bacteria taking thus the energy needed for the vital functions.

DESCRIPTION OF PROCESSES

Denitrification is the process of reducing nitrates (NO_3, NO_2, N_2O, N_2) by denitrification bacteria (anaerobe), which is an endodermic process. Here, one can mention bacteria that reduce nitrates into nitrites: *Bacterium vulgare, Bacterium coli, Bacillus mycoides, Bacillus pyocyaneus*, etc.; and bacteria that reduce nitrites into nitrogen oxides and gas nitrogen: *Pseudomonas denitrificans, Thiobacillus denitrificans, Bacterium denitrificans*, etc.

By denitrification, nitrates are substituted from the oxygen which contributes to the oxidation of the organic matter according to the global scheme (Germon and Couton 1999):

$$5(CH_2O) + 4NO_3 + 4H_3O = 2N_2 + 5CO_2 + 11H_2O$$

Examples of reducing nitrates (Budoi and Penescu 1996):

$2KNO_3 + C = 2KNO_2 + CO_2$;
$2KNO_2 + C = N_2O + K_2CO_3$;
$2N_2O + C = 2N_2 + CO_2$.

The resulted compounds NO_2, N_2O, NO and N_2 are mostly lost in the atmosphere. Thus, denitrification is perceived from the early research (Dehérain and Maquenne 1882) and largely even today (almost by the entire expert literature) as a loss of usable nitrogen for crop plants.

The atmosphere is the main tank of usable nitrogen for plant nutrition through the existence of 70–80,000 t of nitrogen above each hectare of arable land. Once in the soil (by biological fixation, rain and fertilizers) it can be lost (through leaching, denitrification-volatilization and export with the crop) or it can be immobilized in the soil at a biological (by the clay-humic complex) or biochemical level (of microorganisms). In the soil, total nitrogen ranges in general between 0.1–0.3%.

During the last century, at the same time with the production increase of synthesis fertilizers, the intensification of vegetable crops (which biologically fix the molecular nitrogen, between 5–200 kg/ha/year) and return to the soil of the atmosphere nitrogen pollution (with the help of rain, NH_4 and NO_3 between 2–13 kg/ha/year are brought in the soil) resulted from the combustion of fossil fuels, and the N content increases a lot in the biosphere. The result is a general growth of the nitrogen content in the soil-underground and aquifer water, generating a worrying situation in many agricultural areas.

The decrease of nitrates and nitrites in the soil by denitrification at a level of inferior oxides and molecular nitrogen and their release this way into the atmosphere can be considered, therefore, in certain situations and at certain intensities as a protective mechanism (of balance) contrary to their accumulation in the soil-underground and aquifer water. Thus, there are researchers who claim that one can benefit from the functioning of this natural process, run to reduce the nitric pollution or to put into function treatment systems localized where there is nitric pollution (Smith et al. 1976, Mariotti 1994, Hénault and Germon 1995, Hill 1996, Bednarek et al. 2014). Treatment systems by biological denitrification of the residual solid suspensions, which include a biological reactor with automatic control of bacteria and aeration, using methanol as carbon source was even licensed and was a very important process in bioremediation.

One must also mention that nitrification and denitrification lead to the release in the atmosphere of the nitrous oxide (N_2O) gas involved in the increase of the greenhouse effect (Germon et al. 1999). Its direct involvement in the increase of the greenhouse effect is 2% (Mackay and Khalil 1991). Agriculture is responsible almost 50% of N_2O emissions, and these emissions depend less on the intensity of denitrification and more on the number of nitrogen fertilizers used on the soil hydrophysics and on the absence of microorganisms capable of reducing N_2O in N. Emissions vary between 0.2–3% from the nitrogen applied in the soil (Hénault et al. 1998). Stevenson and Grenland (1970) (cited by Avarvarei et al. 1997) estimate that up to 25–30% of the nitrogen applied as fertilizer can be found as N_2O, NO or N_2.

The quantity of N_2O issued from cultivated soil is, in general, less treated by researchers. But it was measured in the river banks where the leaching of nitrates in stagnant and river aquifer water accompanied by conditions favorable to denitrification lead to the high intensity of it. Here Schipper et al. (1993) measured emission of 730 g/ha/day N_2O. A number of potential approaches to increase denitrification on the landscape, thus decrease N export to sensitive coastal systems exist. However, these have not generally been widely tested for their effectiveness at scales required to significantly reduce N export at the whole watershed scale (Seitzinger et al. 2006).

The excessive accumulation of nitrates and nitrites, their translocation in plants and harvests have determined the World Health Organization to establish as admitted dose per human individual, daily consumed during life, without any health risk of 220 mg nitrates and 10 mg nitrites; in the case of children, the limits are less lower (Rusu 2005).

Nitrates (NO_3^-) as such are not toxic for human organisms and animals. Their toxicity is explained by the reduction processes which can take place in the organism with its transformation into NO_2^-.

Nitrites (NO_2^-) entered in the blood prevent the cell oxygenation by blocking the hemoglobin functions, which they transform into methemoglobin. The gravity of the disease is related to the age of the organism and the quantity of hemoglobin blocked.

A possible danger presented by nitrates also comes from the fact that once reduced, nitrites can combine with nitrogen compounds (secondary or third amines, amides and amino acids), present in the gastrointestinal tract and they can form cancer nitrosamines (Mănescu 1984).

As these risks upon organisms are known, there has been almost general concern lately in order to reduce the nitrate content from the soil-underground, aquifer water and the harvest obtained.

CONTROL OF DENITRIFICATION FROM THE SOIL

Factors with direct influence on the denitrification can be considered the following: the existence of denitrification microflora, anoxia (lack of oxygen), availability in oxidized nitrogen and carbonic substrate.

The functioning of the biological denitrification in the soil is ensured by the anaerobic denitrification microflora and optional, by the anaerobic bacteria microflora. So, the level of denitrification populations is rarely a limiting factor of the mechanism of denitrification in the soil. The level of denitrification populations of 106 bacteria/gram of soil are frequent in the arable horizon (Chèneby et al. 1998); these populations decrease in a depth (Weier and McRae 1992) without disappearing completely; the presence of denitrification bacteria has been demonstrated at different depths of aquifer water (Mariotti 1994). In the absence of nitrates and of the conditions favorable for denitrification, these bacteria develop a metabolism adapted to survival conditions until favorable environmental conditions are ensured (Jorgensen and Tiedje 1993).

Oxygen is an inhibitor of the functioning and/or the synthesis of certain enzymes involved in the denitrification process. The functioning of denitrification depends on humidity, which conditions aeration and the level of anoxia. Denitrification increases exponentially starting from a capillary porosity of the soil situated, in general, between 60–65% of the total porosity. The rain regime induces a variation of the humidity in the soil, denitrification having thus a variation in time, which makes it very difficult to measure.

The presence of nitrates or an oxidized form of nitrogen is indispensable for the functioning of denitrification:

$$2NO_3- \rightarrow 2NO_2- \rightarrow 2NO \rightarrow N_2O \rightarrow N_2$$
$$(N^{5+}) \quad (N^{3+}) \quad (N^{2+}) \quad (N^{1+}) \quad (N^0)$$

The availability in the organic substrate is, after aeration, the main adjusting factor of the denitrification in the soil. One can establish existing relations between different carbon forms existing in the soil and the potential or real denitrification activity (Reddy et al. 1982).

Factors with indirect influence on the denitrification in the soil are the climate and cultural ones that interfere and modify the previous ones like rain, the contribution of nitrogenous fertilizers, organic matter, soil tillage system, irrigation, etc.

The idea of intervention on the denitrification in the soil is not recent. Dehérain has recommended ever since 1897 (Dehérain 1897) the use of organic fertilizers in the semifermented state in order to reduce denitrification. And Payne 1990 (Rusu 2005) supports the possibility of selective intervention upon populations and genetically denitrification bacteria, controlling thus the level and forms of reducing nitrates from the soil-underground and aquifer water. Still, at the current state of knowledge in the field, the intervention on the denitrification in the soil is more indirect: by modifying the soil physics and nutrition.

The soil tillage system modifies the biological processes in the soil. This is done by physical, hydrophysical factors and other elements specific to the system.

The classical soil tillage system used in traditional agriculture uses mainly mineral fertilizers in conjunction with repeated energic loosenings leads to the intensification of oxidation processes of organic matter reserves, the nitrification becoming intense and also their leaching on the soil profile.

The NO_3^- ion is not fixed in the soil due to the fact that its negative loads are rejected by the negative loads of the clay-humic complex being submitted to leaching. This is relatively reduced under permanent crops and bigger in the case of crops with a discontinuous covering of the land (the radicular system of the plant controls the nitrates in the soil). Leaching is intensified by the excessive soil works during the 'nude' period of the soil.

Hénault and Germon's research (1995), notice in the soils cultivated according to the classical system fairly reduced denitrification, the repeated soil aeration being the first limiting factor. Here, the nitrogen losses by denitrification are in general 3–10 kg/ha/year. The summer plowing, compared to the autumn one, leads to a higher quantity of nitrates in the soil.

Denitrification increases a lot in the case of irrigated crops where there are more optimal conditions like a higher level of humidity, mineral fertilizers and organic debris. Rolston et al. (1978) measured the loss of nitrogen through denitrification under irrigation conditions at 200 kg/ha in a summer month in California.

The soil conservation systems sustain the minimal soil processing and the increase of the percentage of vegetal debris left and incorporated in the soil to at least 30–50%. The research made (Rusu et al. 2009, Moraru et al. 2015) shows important modifications regarding the state of the soil settlement, the degree of hydric stability and the content of organic carbon in the soil according to the soil tillage system. A high percentage of decomposed vegetal debris, a soil settlement less loosened, a better structuring and consequently an increase of the capillary porosity resulted in a balance between humification and mineralization with the growth of the percentage of organic carbon and at the same time a more intense activity of denitrification. The mulch from the soil surface favors more reduced temperatures and the decrease of amplitudes (denitrification is favored by temperatures around 25°C and nitrification by temperatures of 30–32°C). Theoretically, this would mean more reduced leaching of nitrates.

The nitrification process has a maximum intensity when the values of the apparent density are low, ranging between 1.11–1.15 g/cm^3; the content of nitrates is reduced to the half when the apparent density is higher than 1.4 g/cm^3. It is a well-known fact that minimum soil tillage systems also have consequences like a more stuffed state of settlement (in general specific to the conditions and type of soil).

The correction of the soil reaction, the prevention of soil erosion and the use of fertilizers with a slow release of nitrogen are agronomic measures to control the nitrogen content in the soil. The soil texture has a great influence on the accumulation of nitrogen in the soil. A clay or loamy-clay soil contains, in general, three to four times more nitrogen than a sandy soil has under the same climate conditions.

By Wang et al. (2018), changes in soil denitrification increased exponentially when the rates of synthetic N fertilizer application $\leq$ 250 kg N/ha, but above this threshold, there were no further increases. The responses of soil denitrification to N fertilization were negatively correlated with soil clay content, C:N ratio and bulk density.

Conditions favorable to denitrification can be met under natural or artificial meadows on soils rich in organic matter, especially if these are fertilized (Ryden and Dawson 1982).

These processes of oxidation and reduction of nitrates can be measured by redox potential (Eh). Soils with normal aeration have an Eh ranging between 475–650 mV. Over 650 mV oxidation processes prevail, and under 450 mV reduction processes prevail (Mănescu et al. 1994). In the case of potential lower than 400 mV nitrates disappear from the soil and are reduced to inferior oxides and molecular nitrogen (Parr 1969, Renault et al. 1997). Applying organic matter in large quantities leads the soil to low redox potential, favoring the growth and metabolism of anaerobic microorganisms and stimulating denitrification. Some study suggests that agricultural practices aimed to increase the availability of labile organic matter, such as acetate, are beneficial in buffering reactive N excess in soils and to reduce NO_3- leaching toward groundwater and surface-water (Castaldelli et al. 2019).

Denitrification from the arable layer of the soil is not enough to reduce or annihilate the leaching of nitrates and the avoidance of pollution of aquifer water. Denitrification can also take place during the leaching of nitrates in subarable or at an aquifer level (Mariotti 1994). Starr and Gillham (1993) show that denitrification at these levels is not uniform; it depends on the carbon availability and the depth of the aquifer water. Thus, at a level of almost 1 m of the aquifer water, where there is organic carbon and there are conditions of anaerobiosis, there is also denitrification and the only limiting factor remains thus the nitric nitrogen. At the same time, a soil with the level of aquifer water under 4 m deep, where there is no organic carbon, and the water remains oxygenated and denitrification is absent.

In expert literature, several experimental devices are mentioned with different results, where one artificially tries to stimulate denitrification underground and in aquifer water under situations of excessive pollution with nitrates (Janda et al. 1988, Hamon and Fustec 1991, Schipper and Vodvodic 1998). The experimental devices started by ensuring the organic substrate by introducing sawdust or straws in the trenches, where the aquifer level is superficial, and injecting by drilling the soluble carbonic substrate (methanol or ethanol), where the aquifer water is a higher depth. These experimental devices show that a very detailed technology is needed in order to ensure their efficiency and to avoid collateral, negative effects for the environment.

The most famous artificial device of reducing nitrates through denitrification is the one from Israel, put into function since 1977. The system is designed to treat almost 270,000 m^3/day urban residual water of the city of Tel Aviv, reducing the nitrogen quantities from 7–22 mg/l and stabilizing at 5–7 mg/l, and it can be used at irrigations (Kanarek and Michail 1996).

INFLUENCE OF THE TILLAGE SYSTEM ON THE SOIL PROPERTIES

Data presented in this section was obtained at the University of Agricultural Sciences and Veterinary Medicine in Cluj-Napoca, Romania, within the Research Center for Minimal Systems and Sustainable Agricultural Technologies. The influence of soil tillage system upon soil properties was studied on several soil types (Table 3.2, MESP 1987; SRTS 2003). The experimental soil tillage systems were as follows:

- Classic system: V_1 – classic plough + disc – 2x (witness).
- Minimum tillage systems: V_2 – paraplow + rotary harrow, V_3 – chisel plow + rotary harrow, V_4 – rotary harrow.

Table 3.2 Initial soil properties (0–20 cm) on different soil types at the experimental field

Type of soil (WRB-SR, 1998)	*Clay content, %*	*Humus, %*	*WSA, %*	*pH*	*P.m.m., mm*	*T.m.m, °C*
Chernozem cambic	43.1	3.52	78	6.73	500	8.8
Phaeozem tipic	43.2	3.92	76	6.71	500	8.8
Haplic luvisols	42.0	2.49	65	6.06	613	8.2
Fluvisol molic	41.6	3.01	61	7.25	613	8.2

WSA—Water stabilite macro-aggregation; P.m.m.—Precipitation medium multi-annual; T.m.m.—Temperature medium multi-annual.

To quantify the change in soil properties under different tillage practices, determinations were made for each cultivar (maize – *Zea mays* L., soy-bean – *Glycine hispida* L. Merr., wheat – *Triticum aestivum* L., spring rape – *Brassica napus* L. var. *oleifera* D.C. / potato – *Solanum tuberosum* L.) in four vegetative stages (spring, 5–6 leaves, bean forming and harvest). Soil parameters monitored included soil water content (gravimetric method, Aquaterr probe—Frequency domain reflectometry), soil bulk density (determined by volumetric ring method using the volume of a ring 100 cm^3), soil penetration (using a Fieldscout SC900 Penetrometer), water-stable aggregates, soil permeability (using the Infiltrometer method) and organic matter content. The average result values obtained in the vegetal phases were statistically analyzed using the last four cultivation years within the crop rotation for every type of soil. The results were analyzed using ANOVA and Duncan's test (PoliFact 2010). A significance level of $P \leq 0.05$ was established a priori.

Long-term field experiments provide excellent opportunities to quantify the long-term effects of soil tillage systems on soil properties. The hydrological function of the soil (especially the capacity to retain an optimum water quantity, and then gradually make this available for plant consumption) is one of the most important functions determining soil fertility, productivity and soil evolution. Intrinsic soil properties such as organic matter and texture, along with applied tillage practices combine to modify the soil structure, porosity, permeability and water capacity. This, in turn, is a critical factor in the water cycle and affects water accumulation in the soil.

Statistical analysis of the results showed that the differences in accumulated soil water depended on the variants of soil tillage and type of soil. Soil texture and structure have a strong effect on the available water capacity. The results clearly demonstrate that minimum tillage systems promote increased humus content (0.8–22.1%) (Table 3.3) and increased water stabile aggregate content (1.3–13.6%) at the 0–30 cm depth compared to conventional tillage (Table 3.4).

Table 3.3 The influence of soil tillage system upon organic matter content (OM, %; 0–30 cm)

Type of soil	*Soil tillage systems*	*Classic plow + disc –2x*	*Paraplow + rotary harrow*	*Chisel plow + rotary harrow*	*Rotary harrow*
Chernozem	OM, %	3.51a	3.54a	3.87a	3.61a
cambic	Significance (%)	$^{wt.}$(100)	^{ns}(100.8)	^{ns}(110.2)	^{ns}(102.8)
Phaeozem	OM, %	3.90a	4.13b	3.93ab	3.98ab
Tipic	Significance (%)	$^{wt.}$(100)	*(106.0)	^{ns}(100.9)	^{ns}(102.2)
Haplic	OM, %	2.48a	2.94ab	3.02b	2.82ab
Luvisols	Significance (%)	$^{wt.}$(100)	*(118.6)	*(122.1)	^{ns}(113.9)
Fluvisol	OM, %	3.03a	3.12ab	3.09ab	3.23b
Molic	Significance (%)	$^{wt.}$(100)	^{ns}(103.1)	^{ns}(102.0)	^{ns}(106.5)

Note: OM—organic matter; wt—witness, ns—not significant, *positive significance, a, ab, b, c—Duncan's classification (the same letter within a row indicates that the means are not significantly different.)

Statistical analysis regarding the organic matter content of the studied systems shows significant positive values on Haplic luvisols under paraplow and chisel tillage as well on Typic Phaeozems under paraplow and rotary harrow tillage. Multiple comparisons between systems indicate advantages for using the paraplow on Phaeozems (b), chisel on Haplic luvisols (b) and rotary harrow Molic Fluvisol (b). Multiple analysis of soil classification and tillage system on the hydric stability of soil structure have shown that all variants with minimum tillage are superior (a, b and c) and have a positive influence on soil structure stability.

The increase of organic matter content is due to the vegetal remnants partially incorporated and adequate biological activity in this system. In the case of humus content and also the hydro stability structure, the statistical interpretation of the dates shows an increasing positive significance of the minimum tillage systems application. The soil fertility and wet aggregate stability were initially low, the effect being the conservation of the soil features and also their reconstruction with a positive influence upon the permeability of the soil for water. More aggregated soils permit more water to reach the root zone. This not only increases productivity but it may also reduce runoff and thus erodibility potential.

Table 3.4 The influence of soil tillage system upon water stability of structural macro-aggregates (WSA, %; 0–30 cm)

Type of soil	*Soil tillage systems*	*Classic plow + disc –2x*	*Paraplow + rotary harrow*	*Chisel plow + rotary harrow*	*Rotary harrow*
Chernozem	WSA, %	74.33a	79.00b	78.67ab	80.33b
cambic	Signification (%)	$^{wt.}$(100)	*(106.3)	^{ns}(105.8)	*(108.1)
Phaeozem	WSA, %	80.00a	82.33b	81.00ab	81.67 ab
Tipic	Signification (%)	$^{wt.}$(100)	*(102.9)	^{ns}(101.3)	^{ns}(102.1)
Haplic	WSA, %	63.67a	68.33b	66.67ab	72.33c
Luvisols	Signification (%)	$^{wt.}$(100)	*(107.3)	*(104.7)	**(113.6)
Fluvisol	WSA, %	71.33a	76.00b	75.33b	76.33b
Molic	Signification (%)	$^{wt.}$(100)	*(106.5)	*(105.6)	*(107.0)

Note: WSA—water stability of structural macro-aggregates; wt—witness, ns—not significant, *positive significance, a, ab, b, c—Duncan's classification (the same letter within a row indicates that the means are not significantly different.)

The minimum soil tillage systems and the replacement of ploughing by paraplow, chisel and rotary harrow work minimize soil aeration. The bulk density values at 0–50 cm (Table 3.5) increased by 0–4.7% under minimum tillage systems. This raise was not significant in any of the experimental variants. Multiple comparing and classification of experimental variants align all values on the same level of significance (a).

Table 3.5 The effect of soil tillage system on the bulk density (BD, g/cm^3, 0–50 cm)

Type of soil	*Soil tillage systems*	*Classic plow + disc –2x*	*Paraplow + rotary harrow*	*Chisel plow + rotary harrow*	*Rotary harrow*
Chernozem	BD, g/cm^3	1.32a	1.38a	1.37a	1.36a
cambic	Signification (%)	$^{wt.}$(100)	^{ns}(104.7)	^{ns}(103.9)	^{ns}(103.3)
Phaeozem	BD, g/cm^3	1.22a	1.23a	1.25a	1.22a
Tipic	Signification (%)	$^{wt..}$(100)	^{ns}(100.8)	^{ns}(101.9)	^{ns}(100.0)
Haplic	BD, g/cm^3	1.32a	1.35a	1.34a	1.35a
Luvisols	Signification (%)	$^{wt.}$(100)	^{ns}(102.4)	^{ns}(101.7)	^{ns}(102.4)
Fluvisol	BD, g/cm^3	1.34a	1.34a	1.35a	1.34a
Molic	Signification (%)	$^{wt..}$(100)	^{ns}(100.0)	^{ns}(100.6)	^{ns}(100.0)

Note: BD— bulk density; wt—witness, ns—not significant, a—Duncan's classification (the same letter within a row indicates that the means are not significantly different.)

The soil resistance to penetration, presented as an average of determinations on the four types of soil, shows a stratification tendency of soil profiles within the plow variant, where values are under 1,000 kPa up to the 20–22 cm depth and then suddenly increase over 3,500 kPa below this depth. The significant differences were determined in the minimum tillage systems at 10–20 cm, where the values of resistance to penetration range between 1,500–2,500 kPa. Thus, in the variants worked with minimum tillage system, the soil profile stratification is significantly reduced.

After ten years of applying the same soil tillage system, the data show that soil infiltration and soil water retention are higher when working with paraplow and chisel plow variant with values of 5.54(c) and 5.08(b) $l/m^2/min$, respectively. By contrast, the amount of water retained by traditional tillage was 4.25(a) $l/m^2/min$. The paraplow and chisel plow treatments were more favorable for infiltration and water retention. Positive effects on the saturated hydraulic conductivity of the paraplow (35.7 cm/h) and chisel plow (31.5 cm/h) treated soils were observed compared with the traditional tillage (29.4 cm/h) of the soil.

On haplic Luvisols, a soil with a moderately developed structure and average fertility, the quantity of water accumulated were 1–6% higher under paraplow(b), chisel plow and rotary harrow tillage, compared to conventional tillage (Table 3.6). On molic Fluvisols and cambic Chernozems, soils with good permeability, high fertility and low susceptibility to compaction, accumulated water supply was higher (representing 11–15%) for all minimum soil tillage systems. In the four soils tested, the paraplow was the better at water conservation (as evidenced by multiple comparisons and variants–b, c), showing an increase in the water reserve in the soil of 4.8–12.3%.

Table 3.6 The effect of soil tillage system on the water supply accumulated in soil (W, m^3/ha; 0–50 cm)

Type of soil	*Soil tillage systems*	*Classic plow + disc –2x*	*Paraplow + rotary harrow*	*Chisel plow + rotary harrow*	*Rotary harrow*
Chernozem	W, m^3/ha	936a	1.051b	1.047b	1.039b
cambic	Signification (%)	$^{wt.}$(100)	*(112.3)	*(111.9)	*(111.0)
Phaeozem	W, m^3/ha	842a	882b	875a	859a
Tipic	Signification (%)	$^{wt.}$(100)	*(104.8)	^{ns}(103.9)	^{ns}(102.0)
Haplic	W, m^3/ha	850a	901b	870a	859a
Luvisols	Signification (%)	$^{wt.}$(100)	*(106.0)	^{ns}(102.3)	^{ns}(101.0)
Fluvisol	W, m^3/ha	878a	1.010c	998b	987b
Molic	Signification (%)	$^{wt.}$(100)	*(115.0)	*(113.7)	*(112.4)

Note: W—water supply accumulated in soil; wt—witness, ns—not significant, *positive significance, a, b, c—Duncan's classification (the same letter within a row indicates that the means are not significantly different.)

CONCLUSIONS

The increase of the greenhouse effect, the atmosphere chemistry, the soil and water pollution are current issues that are very publicized and which concern the research from the entire world. The gas transfer in the soil is involved in many environmental problems.

The improvement of soil fertility and the nutrition conditions of crop plants is made through a complex of agronomic measures: soil tillage, fertilizers, amendments, crop rotation, etc. But it is important that, through these measures, one shall adjust consciously the synthesis processes—decomposition and mineralization of the organic substance and immobilizing nitrogen pollution. One cannot talk of a strict delineation of these processes which take place in the soil, but the highlighted aspects are a result of all these with a bigger or smaller contribution of each one of them.

The research during the last years shows the fact that the gas exchange soil-atmosphere and the pollution of the underground and of aquifer water is not only a physical process of transfer, depending on the soil structure and drainage, but also privately related to the microbiological activity in the soil: the dynamics and functioning of the microbial populations, their space distribution, the substrate transfer associated to their functioning and the technology applied.

Reduced tillage systems represent an alternative to conventional tillage. This study demonstrated that increased soil organic matter content, aggregation and permeability are all promoted by minimum tillage systems. The implementation of such practices ensures a greater water reserve even across different soil types. The practice of reduced tillage is ideal for enhancing soil fertility, water holding capacity and reducing erosion. The advantages of minimum soil tillage systems can be used to improve methods in low producing soils with reduced structural stability on sloped fields as well as measures of water and soil conservation on the whole ecosystem.

Presently, it is necessary to make a change concerning the concept of conservation practices and to consider a new approach regarding the control of erosion. The actual soil conservation must be looked upon beyond the traditional understanding of soil erosion. The real soil conservation is represented by carbon management. We need to focus on an upper level concerning conservation by focusing on soil quality. Carbon management is necessary for the complexity of matters, including soil, water management, field productivity, biological fuel and climatic change.

One must also mention that while in the soil these processes are relatively easy to control, in the underground and in aquifer water their control remains very difficult. The artificial devices to intensify denitrification in the places with nitric pollution still need research to set up an adequate, efficient technique, which doesn't have other negative effects upon the environment. Therefore, it results in certainty that it is a lot easier to prevent nitric pollution than to restore the quality of the underground and aquifer water.

REFERENCES

Avarvarei, I., Davidescu, V., Mocanu, R., Goian, M., Caramete, C. and Rusu, M. (eds) 1997. Agrochemistry. Sitech, Craiova, p. 652 (in Romanian).

Bednareka, A., Szklareka, S. and Zalewski, M. 2014. Nitrogen pollution removal from areas of intensive farming–comparison of various denitrification biotechnologies. Ecohydrology and Hydrobiology. 14(2): 132–141. DOI:10.1016/j.ecohyd.2014.01.005

Berca, M. (ed.) 2011. Agrotechnics: the modern transformation of agriculture. Ceres, Bucuresti (in Romanian).

Blaga, Gh., Rusu, I., Udrescu, S. and Vasile, D. (eds) 1996. Pedology. Didactics and Pedagogy, Bucuresti (in Romanian).

Budoi, Gh. and Penescu, A. (eds) 1996. Agrotechnics. Ceres, Bucuresti, p. 439 (in Romanian).

Castaldelli, G., Colombani, N., Soana, E., Vincenzi, F., Fano, E.A. and Mastrocicco, M. 2019. Reactive nitrogen losses via denitrification assessed in saturated agricultural soils. Geoderma. 337: 91–98, DOI:10.1016/j.geoderma.2018.09.018

Chèneby, D., Hartmann, A., Henault, C., Topp, E. and Germon, J.C. 1998. Diversity of denitrifying microflora and ability to reduce N_2O in two soils. Biology and Fertility of Soil. 28: 19–26.

Dehérain, PP. and Maquenne, L. 1882. Sur la reduction des nitrates dans la terre arable. C.R. l'Académie des Sciences, Paris. 95: 691–693.

Dehérain, P.P. 1897. La reduction des nitrates dans la terre arable. C.R. l'Académie des Sciences, Paris. 124: 269–273.

Florea, N. (ed.) 2003. Degradation, Protection and Improvement of Soils and Lands. University, Bucureşti (in Romanian).

Germon, J.C. and Couton, Z. 1999. La dénitrification dans les sols: Regulation de son fonctionnement et applications a la depollution. Le Courrier de l'environnement de l'INRA. 38: 67–74.

Germon, J.C., Henault, C., Garrido, F. and Reau, R. 1999. Mecanismes de production, regulation et possibilites de limitation des emissions de N_2O a l'echelle agronomique. C.R. Academy Agricultural, Paris. 85: 148–162.

Hamon, M. and Fustec, E. 1991. Laboratory and field study of an in situ grounwater denitrification reactor. Journal of Water Pollution Control Federation. 63: 942–949.

Hénault, C. and Germon, J.C. 1995. Quantification de la dénitrificationet des emissions de protoxyde d'ayote par les sols. Agronomie. 15: 321–355.

Hénault, C., Devis, X., Page, S., Justes, E., Reau, R. and Germon, J.C. 1998. Nitrous oxide emission under different soil and land management conditions. Biology Fertility of Soils. 26: 199–207.

Hill, A.R. 1996. Nitrate removal in stream riparian zones. Journal Environmental Quality. 25: 743–755.

Janda, V., Rudovsky, J., Wanner, J. and Mahra, K. 1988. In situ denitrification of drinking water. Water Science Technology. 20: 215–219.

Jorgensen, K.S. and Tiedje, J.M 1993. Survival of denitrifiers in nitrate-free anaerobic environments. Applied Environmental Microbiology. 59: 3297–3305.

Kanarek, A. and Michail, M. 1996. Groundwater recharge with municipal effluent: Dan Region Project, Israel. Water Science Technology. 34: 227–233.

Mackay, R. and Khalil, M.A.K. 1991. Theory and development of a one dimensional time dependent radiative convective climate model. Chemosphere. 22(3–4): 383–417. DOI: 10.1016/0045-6535(91)90326-9

Mariotti, A. 1994. Denitrification in situ dans les eaux souterraines: Processus naturels ou provoques: Une revue. Hydrogeology. 3: 43–68.

Mănescu, S. (ed.) 1984. Hygiene Treaty. Medicală, Bucureşti (in Romanian).

Mănescu, S., Cucu, M. and Diaconescu, M.L. (eds) 1994. Environmental Health Chemistry. Medicală, Bucureşti (in Romanian).

Moraru, P.I., Rusu, T., Guş, P., Bogdan, I. and Pop, A.I. 2015. The role of minimum tillage in protecting environmental resources of the Transylvanian Plain, Romania. Romanian Agricultural Research. 32: 127–135.

Parr, J.F. 1969. Nature and significance of inorganic transformations in tile-drained soils. Soils Fertility. 32: 411.

Reddy, N.R., Sathe, S.K. and Salunkhe, D.K. 1982. Phytates in legumes and cereals. Advances Food Research. 28: 1–92.

Renault, P., Parry, S., Sierra, J. and Bidel, L. 1997. Transferts de gaz dans les sols. Courrier de l'Environnement de l'INRA. 32: 33–50.

Rolston, D.E., Hoffman, D.L. and Toy, D.W. 1978. Field measurement of denitrification: I. Flux of N_2 and N_2O. Soil Science Society America Journal. 42: 863–869.

Rusu, T. (ed.) 2005. Agrotechnics. Risoprint, Cluj-Napoca, p. 334.

Rusu, T., Gus, P., Bogdan, I., Moraru, P.I., Pop, A.I., Clapa, D., Marin, D.I., Oroian, I. and Pop, L.I. 2009. Implications of minimum tillage systems on sustainability of agricultural production and soil conservation. Journal of Food Agriculture and Environment. 7(2): 335–338.

Ryden, J.C. and Dawson, K.P. 1982. Evaluation of the acetylene inhibition technique for field measurements of denitrification in grassland soils. Journal Science Food Agriculture. 33: 1197–1206.

Schipper, L.A., Cooper, A.B., Harfoot, C.G. and Dyck, D.J. 1993. Regulators of denitrification in an organic riparian soil. Soil Biology Biochemistry. 25: 925–933.

Schipper, L.A. and Vodvodic, M.V. 1998. Nitrate removal from groundwater technique for the measurement of denitrification in grassland soils. Journal Science Food Agriculture. 33: 1197–1206.

Seitzinger, S., Harrison, J.A., Bohlke, J.K. Bouwman, A.F., Lowrance, R., Peterson, B., Tobias. C. and Drecht, G.V. 2006. Denitrification across landscapes and waterscapes: A synthesis. Ecological Applications. 16: 2064–2090. doi.org/10.1890/1051-0761(2006) 016[2064:DALAWA]2.0.CO;2.

Smith, J.H., Gilbert, R.G. and Miller, J.B. 1976. Redox potential and denitrification in a cropped potatoprocessung waste water disposal field. Journal Environmental Quality. 5: 397–399.

Starr, R.C. and Gillham, R.W. 1993. Denitrification and organic carbon availability in two aquifers. Ground Water. 31: 934–947.

Wang, J., David, R., Chadwick, Y.C. and Xiaoyuan, Y. 2018. Global analysis of agricultural soil denitrification in response to fertilizer nitrogen. Science of the Total Environment. 616–617: 908–917. Doi: 10.1016/j.scitotenv.2017.10.229.

Weier, K.L. and MacRae, I.C. 1992. Denitrifying bacteria in the profile of a brigalow clay soil beneath a permanent pasture and a cultivated crop. Soil Biology Biochemistry. 24: 919–923.

***ICPA, 2005. Code of Good Agricultural Practice. Bucuresti, Vol. 1–3 (in Romanian).

***MESP, 1987. Podologic Studies Elaboration Metodology. Pedologic and Agrochemical Ins., Bucuresti. Vol, 1–3 (in Romanian).

***PoliFact, 2010. ANOVA and Duncan's Test PC Program for Variant Analyses Made for Completely Randomized Polifactorial Experiences. USAMV Cluj-Napoca.

***SRTS, 2003. Romanian System of Soil Taxonomy. (ed.) Estfalia, Bucuresti, p. 182 (in Romanian).

***WRB-SR, 1998. World Reference Base for Soil Resources. World Soil Resources Report 84. ISSS, ISRIC.

Chapter 4

Shelterbelt as Significant Element of the Landscape for the Changes of Organic Nitrogen Compounds in Groundwater and Soil

Lech Wojciech Szajdak

Institute for Agricultural and Forest Environment,
Polish Academy of Sciences, Poznan, Poland.

INTRODUCTION

Increasing recognition that intensive agriculture introduces many chemical compounds of well-known and unknown structures and impoverishments of soil organic matter to groundwater indicates that it is necessary to change attitudes about the phenomena taking place in the agricultural landscape. We should stop thinking of each process in agricultural development as being isolated from the whole environmental, economic and/or social context. Therefore, the increased recognition of the natural processes in question facilitates understanding and improved management of various environmental threats to rural areas caused by agricultural intensification. Another problem of importance for the development of sustainable agriculture is the conservation of organic matter resources. The recent achievements in understanding primary production rates in various ecosystems allow us to evaluate more precisely the balance of organic matter in the soil of cultivated fields. These results can contribute to the development of programs

---

For Correspondence: E-mail: lech.szajdak@isrl.poznan.pl

optimizing agricultural production rates with that of organic matter conservation, which is important for sustainable farming (Ryszkowski 1989).

Organic matter has a significant influence on the mobility and fate of inorganic and organic compounds of low and high molecular weight. Interactions of organic matter, metals, phosphorus and organic substances in soils may lead to the formation of stable complexes. Organic substances in the soil tend to react with inorganic constituents, but the mechanism of the reaction and the properties of the resulting complexes have not been fully clarified. The metal complexing ability of humic substances arises primarily from their carboxyl, phenolic hydroxyl and carbonyl groups (Hayes and Swift 1978, Stevenson 1982), but other functional groups present in smaller amounts may have the ability to form very strong complexes under certain circumstances, e.g., small amounts of porphyrin groups have been identified (Table 4.1) (Figures 4.1–4.3).

Table 4.1 Functional organic groups as ligands in humic substances

Strong		***Weak***	
Primary amines	$R{-}\ddot{N}H_2$	Carbonyl groups	$R_2C{=}O$
Secondary amines	$R_2\ddot{N}H$	Etheric groups	R_2O
Tertiary amines	$R_2\ddot{N}{-}R$	Ester groups	$R{-}C({=}O){-}O{-}R$
Aromatic amines	$=N{:}$	Amidic groups	$R{-}C({=}O){-}NH{-}R$
Carboxylic anions	$R{-}C({=}O){-}O^-$	Sulfidic groups	$R{-}S{-}R$
		Hydroxyl groups	$R{-}O{-}H$
Enolic anions	$-C{=}C{-}O^-$	Sulfoxide	$R{-}S({=}O){-}R$
Alkoxy anions	$R{-}O^-$		
Phenoxy anions	$C{-}O^-$		
Mercapto anions	$R{-}S^-$		
Phosphate anions	$R{-}O{-}P({=}O)(O^-){-}O^-$		
Phosphonic anions	$R{-}CH_2{-}P({=}O)(O^-){-}O^-$		
Sulfinic anions	$R{-}S({=}O){-}O^-$		
Sulfonic anions	$R{-}S({=}O)_2{-}O^-$		

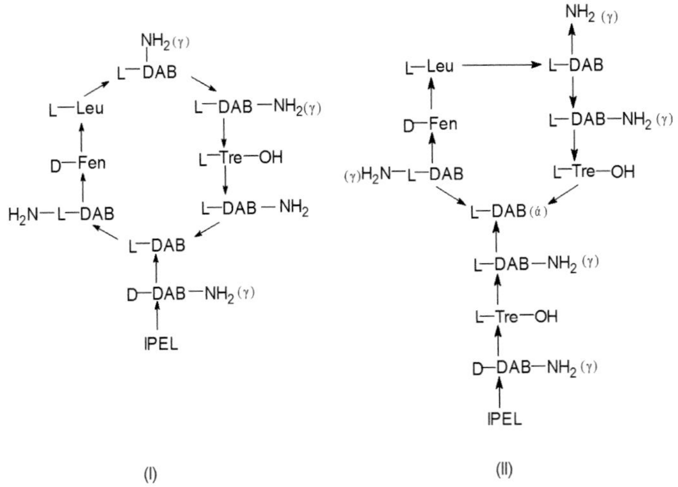

(I) (II)

Figure 4.1 Polymixine B—the product of *Bacillus polymyxa strains* in groundwater. Structure of polimyxine B from eight (I) or seven (II) of amino acids.
(where: bond –CO–NH–, Tre—treonine, DAB—L–α,γ-diaminebutyric acid, Fen—phenylalanine, Leu—leucine, IPEL—isopelargonic acid)

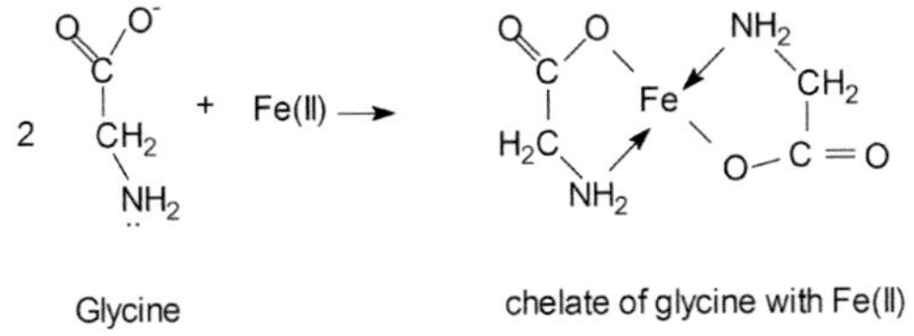

Glycine chelate of glycine with Fe(II)

Figure 4.2 Chelate of glycine with iron ion (II).

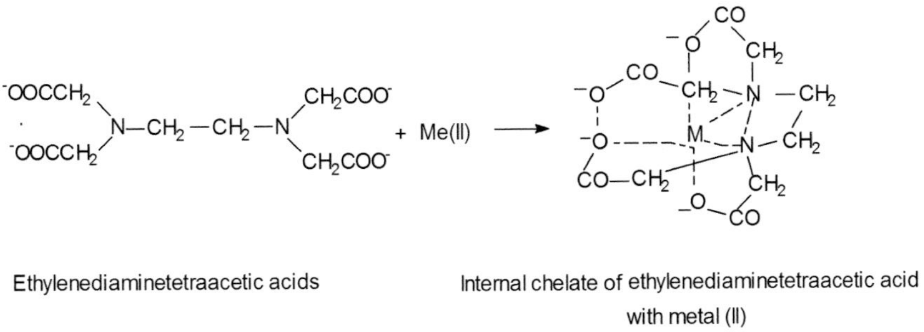

Ethylenediaminetetraacetic acids Internal chelate of ethylenediaminetetraacetic acid with metal (II)

Figure 4.3 Formation of an internal chelate of ethylenediaminetetraacetic acid with metal (II).

Metal-organic complexes contain micro and macro elements in which humus compounds of well-known and unknown structures such as amino acids represent strong organic chelate ligands (Stevenson 1982, Fitch and Stevenson 1984). Therefore, the dissolved organic substances participate in the conversions of organic matter and transport of the inorganic and organic substances throughout the groundwater.

Earlier studies have shown a high concentration of dissolved chemical substances in the groundwater of agricultural landscape where light soils dominate (Szpakowska and Życzyńska-Bałoniak 1994a), especially organic compounds, in which humic substances dominate (Życzyńska-Bałoniak et al. 1998).

The efficient elements of agricultural landscape limiting the migration of nitrogen organic and inorganic substances in groundwater are the biogeochemical barriers in the form of shelterbelts, meadows, peatlands, woodlots, bushes, wetlands and semi-natural plant association, which separate agricultural fields from the watercourses (Szpakowska and Życzyńska-Bałoniak 1994b, Życzyńska-Bałoniak et al. 1993, 1996). Significant concentrations of nitrogen organic compounds from groundwater are simultaneously taken by the roots of trees and supplied into the soils by root exudates (Vancura 1967, Claudius and Merhotka 1973, Smith 1976, Peterson 1983, Luckner 1990),

There is scant information available concerning the impact of biogeochemical barriers on the limitation of amino acid migration throughout groundwater in the agricultural landscape. In our earlier study, we dealt with the problem of the occurrence of amino acids in the water of small ponds and their sediments. In this study, it was shown that a high content of amino acids is included in the sediments of small ponds. In addition to this, we observed that neutral amino acids dominated in all groups of amino acids in the groundwater (from 44 to 61%). Basic amino acids were shown to have the highest content in sediments (from 60 to 76%) (Życzyńska-Bałoniak et al. 1996).

The aim of this study was twofold: 1) to show the migration of amino acids from the adjoining cultivated fields and transport them with groundwater and 2) to estimate the limited degree of shelterbelts as a biogeochemical barrier for the spread of pollution in the form of free and bound amino acids dissolved in the groundwater that is present in the soil.

The high concentration of organic and inorganic compounds migrating among ecosystems of the agricultural landscape in groundwater leads to the degradation of water quality.

These chemicals are known as nonpoint pollutions. Degradation of water quality by nonpoint source inputs is not a recent phenomenon, but the visibility of its effects has increased because of greater regulatory, public and research awareness of this pollution source. More recently, the impact of various forms of nitrogen, phosphorus, sulfur, oxygen demand, microbial species and other materials on nonpoint source pollution has received attention. This particular focus coincides with the success in reducing point source inputs to receiving waters, resulting in a greater emphasis on evaluation and control of nonpoint sources. Thus, control of nonpoint sources pertains to dealing with and requiring information on the phenomena resulting from rainfall-runoff and other diverse water movement processes responsible for transporting pollution. The modeling of nonpoint pollution serves as a guide for understanding and quantifying the various soil, vegetation and climatic elements responsible for controlling water quality. That is by the conceptual structuring of the processes involved between land use and the ultimate impact on receiving water quality, modeling clarifies data and research needs. The phenomena of chemical, biochemical and biological material transported from rainfall-runoff to receiving waters consist of two broad areas of research (Overcash and Davidson 1980, Overcash et al. 1980, Reddy and Rao 1983):

(i) Transformations in the form and amount of material present at the land surface.
(ii) The transfer and transport of material from the land surface into water moving across or throughout the land and ultimately to receiving waters (Overcash and Davidson 1980, Overcash et al. 1981).

The impact of agricultural activities on water quality results from a complex set of circumstances. Source control is an attempt to reduce the edge-of-field loss of pollutants. Delivery control is focused on measures that prevent pollutants' movement from the field to the aquifer stream. Agricultural pollutants (organic and inorganic) can be considered as two major groups: land-based and management-related.

- The land-based pollutants are associated with the soil and result from erosion and subsequent movement of soil particles to surface waters.
- Management-related pollutants are those that are applied to the crop or land to enhance the productivity of cultivated plants and reduce pests.

Some of the pollutants may be soluble or suspend in water and move with the runoff. Others may adsorb to the soil and move as the soil moves when eroded (Loehr 1984).

Nonpoint pollutions represent a broad group of inorganic and organic compounds of well-known and unknown structure: nitrates, nitrites, phosphates, heavy metals (antimony, arsenic, beryllium, boron, cadmium, chromium, cobalt, copper, germanium, lead, nickel, manganese, mercury, molybdenum, selenium, silver, zinc, vanadium, yttrium), cyanide, fluorine, sulfide ions, organic nitrogenous compounds (amino acids, proteins, peptides, amines, alkaloids, antibiotics, creatine and creatinine), carbohydrates (reducing sugars, starch, soluble carbohydrates, cellulose, holocellulose, α-cellulose, hemicellulose), vitamins, crude fibre, fatty acids and lipids, flavonoids, and related compounds (lignin, phenolic compounds including phenolic acids), plant pigments (including chlorophyll and carotenoids), sterols, pesticides (including herbicides, carbamates, polychlorinated biphenyl), detergents, anionic surfactants, humics and resistant residues and suspended matter consisting of plant and animal origins.

Numerous contaminants are anthropogenic, including petroleum products, chlorinated solvents and pesticides, but nitrate contamination can be either natural or man-made and is one of the most common chemical contaminants of groundwater. Nitrates represent great threats to rural areas. In general, nitrate concentrations are highest in groundwater nearest the land surface when nitrogen sources are present (Hallberg and Keeney 1993).

Estimation of individual organic compounds is not often necessary in ecological investigations, but if needed it usually relates to a specific problem that may have to be investigated in depth. For example, the relationship of humus composition to different eco-habitats or the influences of organic decomposition products on mineral nutrients are problems that may require detailed organic analyzes (Allen 1974).

The overall objective of research studying the problem of agricultural nonpoint source pollution is to reduce the hazard associated with sediment and agricultural losses. There are several key processes that must be estimated, such as infiltration, runoff, erosion, adsorption-desorption and chemical transformation in the soil. Source control is an attempt to reduce the edge-of-field loss of pollutants. Delivery control has focused on measures that prevent pollutant movement from field to stream or aquifer (Brum et al. 1968).

The most significant factors that determine chemical degradation rates in the agricultural landscape are soil type, soil water content, pH, temperature, clay and organic matter content. Increasing the soil pH will generally increase the rate of chemical degradation of many compounds (Goring 1967, Helling et al. 1971, Sewell and Alphin 1975, Enfield et al. 1981, Reddy 1982, 1983, Sharpley and Smith 1992). The most profound and yet unpredictable influence on chemical degradation rates is caused by the soil microbiological populations and the soil environmental variables that control their activity. Temperature and soil-water content are two environmental factors that have been most intensively studied in this regard. Increasing the temperature will significantly increase the solubility of chemicals and the rate of their conversions (Frost and Pearson 1961, Lasaga et al. 1981, Connors 1990).

In addition to this, increased microbial activity and decreased adsorption associated with higher temperature generally enhance chemical degradation. The half-lives of chemicals are known to increase with decreasing soil-water content (Walker 1987, Clark and Gilmour 1983, Walker and Hollis 1994, Walker et al. 1995, Walker et al. 1996, Eggen and Majcherczyk 2006).

Higher adsorption and lower microbial activities may be responsible for reduced degradation as the soil-water content is decreased. At high soil-water content (approaching saturation), the chemical degradation rate is determined by the relative rates of decomposition under aerobic versus anaerobic conditions. However, compounds in a highly oxidized state tend to resist microbiological degradation under aerobic conditions but are susceptible to reductive decomposition under anaerobic conditions (Goring and Lindahl 1975, Reddy et al. 1980a, b, c, d, Khaleel et al. 1981).

Chemical compound disappearance under field conditions may be the result of conversions, degradation, volatilization, evolution and in some cases formation of bound residues (Szajdak 1996, Szajdak et al. 2003a, b). Transport of organic matter is important because of its large surface area, which has a high potential for transporting adsorbed chemicals. Organic matter is usually enriched in eroded sediment with observed ratios ranging from 1.2 to 4.4. (Crawford and Donigian 1973, Donigian and Crawford 1976, Donigian et al. 1977).

Humic and fulvic acids representing the biggest fractions of organic matter are the most important drivers of many heavy metals in groundwater and soil, which may form complexes leading to the solubilization of heavy metals. The negative aspects of these complexes are their stability and resistance to microbiological degradation in the environment (Figures 4.2–4.4) and are, therefore, accumulated in high concentrations.

The function of metal ions in biochemistry may be correlated with their coordination tendencies. For example, sodium and potassium, which only enter into a weak bonding with typical ligands, chiefly through electrostatic effects, are involved with control processes involving ion transfer. However, Mg(II) and Ca(II) also are largely electrostatic in their interactions but are better complex formers compared to that of sodium and potassium. They exhibit rapid ligand exchange and often are components of enzymes involved in trigger reactions and hydrolysis processes. In addition to that, electronegative ligands such as oxygen donors are favored by both of these groups. Stronger bonding, preference for nitrogen donors and stronger interactions with sulfur donors are found with the ions of Zn, Co, Mn, Fe and Cu. These are also common in enzyme systems that control a variety of these processes. Those that readily form two different oxidation states, especially Cu and Fe, often are involved in redox processes. In addition to this, it is well established that strong interactions between humic substances and metal ions take place because humic substances reveal a high capacity for binding metal ions, forming a potential reservoir of trace metals in groundwater and soils (Bailey et al. 1978).

Although some organic matter is removed as individual pieces having a low density, much of the organic matter is absorbed on fine (clay) soil particles in the form of long-chain carbon molecules. Consequently, organic matter is contained in aggregates and when aggregates are deposited, organic matter, in turn, is deposited (Khaleel et al. 1979a, b, 1980, 1981).

The simplification of the structure of agro-ecosystems connected with the intensification of agricultural production increases the hazards of leaching, wind and water erosion and volatilization of chemical substances from soil (Pearson 1996, Ryszkowski et al. 2002).

Long-term investigations carried out in the Wielkopolska region (West-Polish Lowland) by the Institute for Agricultural and Forest Environment of the Polish Academy of Sciences in Poznań have revealed a high content of chemical and biochemical compounds leaching with groundwater from adjoining cultivated fields (Szajdak and Matuszewska 2000, Szajdak and Meysner 2002a, Szajdak et al. 2002a, b, Szajdak et al. 2003b). Due to their high mobility in the soil profile, they simultaneously leach from soil into groundwater. Some inorganic ions migrate apparently in the form of mineral salts and also mineral-organic complexes (Boratyński 1981, Życzyńska-Bałoniak and Szpakowska 1989, Szajdak et al. 2009). The subsequent decrease of their concentrations into the sorptive complex due to excessive leaching would appear to be a negative effect. They are available and indispensable nutrients for plants and also definitely and significantly participate in the biological, physical, chemical and biochemical processes in soils (Szajdak 1996).

The dispersal of organic and mineral substances including pesticides between ecosystems of the agricultural landscape is difficult to limit. It can, however, be efficiently decreased throughout the system of biogeochemical barriers established in the form of shelterbelts, meadows and peatlands. Therefore, it can be stated that shelterbelts are involved in the migration of chemical and biochemical substances of unknown and well-known structures passing among the ecosystems in the agricultural landscape throughout groundwater or in the adsorbing and adsorbing of such components.

Shelterbelts (mid-field rows of tree afforestation) are a spectacular example of preventing the degradation of the agricultural landscape. These elements of the landscape were introduced and popularized in the Poland Wielkopolska region in the nineteenth century by Dezydery Chłapowski (Figure 4.4).

(a)

(b)

Figure 4.4 System of shelterbelts in Turew.

Shelterbelts have proven to be efficient in protecting the agricultural landscape, including wind protection. Slower wind velocities lead to slower rates of water evapotranspiration and improve the microclimate for agricultural production (Correll 1997). Shelterbelts show the following several positive functions in the agricultural landscape; they restrain water erosion, especially surface erosion, and form very efficient (and effective) biogeochemical barriers limiting the migration of biochemical and chemical substances, including organic and inorganic fertilizers and pesticides running off cultivated fields throughout groundwater. Therefore, they play an important function limiting the spread of what is known as 'nonpoint source pollution'. The mechanism of purification by these landscape elements is a complex one and depends on the distribution, density and the activity of root systems of plants taking nutrients from the soil solution and the sorptive properties of the soil. In addition to this, chemical and biochemical processes occurring in soils are responsible for the decline in the concentration of all biogenic components in groundwater.

Several theories have been proposed to explain why the amounts of chemicals decrease in the groundwater flowing away from adjoining cultivated fields and passing the shelterbelts. One explanation dealt with the problem of a well-developed network of tree roots. They retain more groundwater in their range than cultivated crop roots. Due to their 34% greater water transpiration rate as compared to cultivated plants, trees may also uptake chemical and biochemical compounds more strongly, thereby impacting on the contents of biochemical and chemical substances in the groundwater. However, acting as a kind of lift and force pump, tree roots take up mineral and low-molecular-weight organic compounds from their vicinity. There are these complex processes under oxygen and moisture conditions with the proper redox potential and enzyme involvement, giving rise to humus substances of a well-developed structure.

In addition to that, the movement of ions through groundwater may be limited or accelerated by the physical and chemical properties of a given soil. These properties lead to the accumulation of certain chemicals in the soil and further their absorption by plants. An important function here is also played by the sorptive complex, representing the colloidal part of the solid-state in soil, together with adsorbed exchangeable ions. The soil colloids that form the sorptive complex are carriers of a negative charge, giving the complex a certain cation exchange capacity. An important role in cation exchange sorption is played by humus and clay minerals, although the sorptive capacity of the former significantly exceeds that of the latter. This is due to the greater specific surface area of humus, which adsorbs significant quantities of hydrogen and ammonium ions as well as cations of magnesium, calcium and others. Exchangeable cations adsorbed by humus are significantly more easily evolved from the solid-state of the soil. Therefore, they accelerate their mobility as compared to cation adsorbed by the mineral part of the sorptive complex.

In addition to this, denitrification is one of the mechanisms involved in the conversion of nitrogen in soils under shelterbelts which reduce the nitrates to nitrites leading to the evolution of NO, N_2O and N_2 into the atmosphere. The denitrification in soils under shelterbelts process proceeds faster than in soil under adjoining cultivated fields due to the higher content of organic matter in soils under shelterbelts.

In addition to that, a higher content of organic matter in soils under shelterbelt results in higher moisture concentrations. Both of these factors (high content of organic matter and moisture concentrations) give rise to more favorable conditions for the process of denitrification in shelterbelt soils. However, one negative aspect of this process is the significant loss of nitrogen from soils under shelterbelts.

Shelterbelts significantly decrease the concentration of chemical compounds in groundwater and therefore improve the water quality in the agricultural landscape. Thus, the idea that inspired Dezydery Chłapowski in the nineteenth century has gained an unexpected new significance in the area of intensive agriculture.

MATERIALS AND METHODS

The study was carried out in the shelterbelt located in the Turew Agro-ecological Landscape Park (40 km south of Poznań, West Polish Lowland) (Figure 4.4), where intensive agriculture is observed. Here, cultivated fields represent 70%, meadows 12% and shelterbelts 14% respectively of the Park. Characteristic features of this landscape are shelterbelts created in the nineteenth century by General Dezydery Chłapowski, where shelterbelt and adjoining cultivated fields were introduced on Hapludalfs soils (according to FAO classification).

In respect to this particular environment, the shelterbelt consists mainly of *Robinia pseudacacia* and a small admixture of *Quercus robur and Larix deciduas*, some 30 m wide and 400 m long. Soil samples were taken from 10 loci in the middle of shelterbelt areas and from 10 such of each adjoining cultivated field located 100 m from the shelterbelts, taking the upper 20 cm of soils (humus horizon) during the period of intensive plant growth. Soil samples were air-dried and crushed to pass a 1-mm-mesh sieve and the 10 sub-samples subsequently were mixed so as to prepare a 'mean sample'. These 'mean samples' of the soils under study, in turn, were taken for soil characterization and extraction of HAs.

Mean rainfall was 790 nm, and the mean year temperature was equal to 9.5°C. The groundwater passing throughout the 16.5 m wide shelterbelt was uptaken from the piezometers. The water was filtered by means of a Whatman GT/C filter paper and pH, dry mass, organic and mineral carbon and the forms of nitrogen and dissolved humus substances were determined. Soil samples were taken near the piezometers from the layer at 0–20 depth. Soils then were air-dried and sieved, using a 1 mm sieve. Organic carbon was measured using TOC 5050 analyzes (Shimadzu, Japan) and the content of organic matter was calculated by multiplying the concentration of organic carbon by factor 1.724 (Lityński et al. 1976).

Dry mass was obtained by the evaporation of water and dried at a temperature of 105°C to the constant of weight. The following forms of nitrogen were analyzed: total, organic, ammonium and nitrate according to Hermanowicz et al. (1976). Humic substances were isolated from the water before acidification by hydrochloric acid to pH = 2 and next by passing nonionic resin Amberlit XAD–2 throughout the column (Życzyńska-Bałoniak and Szpakowska 1989). Total nitrogen in soils was determined by sulfuric acid mineralization. Amino acids concentrations in the dry mass of groundwater, humus substances dissolved in water and humic acids extracted from soils were determined using an amino acids analyzer AAA T 339 (Mikrotechna, Prague) (Szajdak and Österberg 1996, Życzyńska-Bałoniak et al. 1996). The activity of urease in soils was measured according to the Hoffmann-Teicher colorimetric method (Szajdak and Matuszewska 2000).

RESULTS AND DISCUSSION

The content of dissolved substances in groundwater passing from agriculture field to leaf afforestation in dry mass was 2,878 mg/l and the mean of year content was equal to 2,074.0 mg/l. An increase in high concentrations of organic substances to

61.03 mg/l was also observed but yearly mean the content was only 43.34 mg/l. Among all organic compounds, humus substances predominated and represented 88% of organic carbon (Table 4.1). The mean of the total nitrogen in the groundwater carried away to the afforestation was 15.0 mg/l but in this value 7.03 mg/l was included in organic forms.

It was observed that during flows of groundwater across leaf afforestation, the content significantly decreased in all of the chemical compounds to 65%, and at the same time, there was a decrease of dissolved organic compounds of 55%, humic substances 80%, but organic nitrogen compounds measured only 23% (Table 4.2).

Table 4.2 Means and averages (*italic*) of physicochemical parameters for the groundwater flows to the shelterbelt from adjoining cultivation field and flows away of 16.5 m wide of the edge in 1997 in mg/l

Physicochemical Parameters of the Groundwater	*The Groundwater Flows Into the Shelterbelt*	*The Groundwater After the Flow of* 17 m *of the Shelterbelt*
pH	*7.29–7.79*	*7.03–7.61*
Dry mass	2074.00 *911.2–2878.40*	729.40 *528.4–1048.8*
Total organic compounds	43.34 *31.20–61.03*	19.45 *6.98–45.19*
Humus substances	38.24 *28.03–50.22*	11.64 *3.42–23.94*
Organic nitrogen	7.03 *0.00–13.83*	5.43 0.00–10.08
Total nitrogen	15.00 *2.99–28.22*	14.42 *3.19–22.43*

The decrease of the concentration of the main dissolved organic compounds in groundwater may indicate a strong process of mineralization in this water and the significant sorption capacity of this soil. At the same time, the soils near piezometers were investigated, from which groundwater was taken for the purposes of research. The respective pH values of soils from the boundary between the field and the afforestation ranged from 5.64–6.12. The mean content of total nitrogen in this soil was 169.19 mg/100 g. However, the pH in the soil under afforestation at a 16.5 distance from the edge of afforestation was smaller and ranged from 4.11 to 4.97, while the mean content of nitrogen was similar to that determined on the boundary between the field and afforestation, measuring 170.36 mg/100 g of soils (Table 4.2).

In both studies, the soil activity of urease was measured. This enzyme participates in the conversion of urea, which is one of the final products created during the degradation of organic nitrogen substances in soil. A decrease in the activity of this enzyme was observed with respect to the distance from the edge of the afforestation. At the same time, the increase of $N–NO_3$ content in groundwater of the afforestation was measured. The study year belonged to that of wet and warm years. These phenomena were accompanied by a strong litter degradation process, which caused the liberation of a higher input of nitrogen to the groundwater (Ryszkowski et al. 1998) (Table 4.3).

Table 4.3 pH and means of the year and averages (*italic*) of the urease activity and the total content of nitrogen in soil

Physicochemical Properties of Soil	*Soil of the Border Between the Field and the Shelterbelt*	*Soil Under the Shelterbelt 16.5 m From the Edge*
pH	*5.64–6.12*	*4.11–4.94*
Urease $mmol^{-1} \cdot g^{-1} \cdot h^{-1}$	6.365 *2.201–10.552*	2.471 *0.612–6.272*
Total nitrogen in mg/100 g of soil	169.19 *121.00–201.05*	170.36 *153.45–189.30*

Humus is composed of 20 to 60 % humic acids (HAs). From twenty to forty-five percent of the nitrogen associated with HA may consist of amino acids or peptides connected to the central core by hydrogen bonds (Harworth 1971). Considerable quantities of amino acids occur in soil in the protein fraction bound to humus, mostly to HAs (Szajdak and Österberg 1996). Humus has a protective effect on the protein complex, preventing its further decomposition (Trojanowski 1973), its protein fraction included in the organic colloid component of soils. The colloidal character of peat soils is stronger than in most mineral soils and the specific surface area of organic colloids is from 2 to 4 times greater than that of montmorillonite minerals (Buckman and Brady 1971, Kwak et al. 1986). A characteristic feature of protein is the high content of various functional groups ($-NH_2$, $=NH$, $-SH$, $-COOH$) as a result of which they possess ion-exchange and complexing properties (Fitch and Stevenson 1984). They can transport complexes of heavy metals and biologically active substances for plants and soil organisms (Schnitzer and Khan 1978).

The groundwater that passed from the cultivated field to the shelterbelt contained 133.49 μg/l of totally free and bound amino acids. The following amino acids dominated: cystathionine (16.33 μg/l), citrulline (14.87 μg/l), histidine (14.30 μg/l), 1-methylhistidine (12.09 μg/l), arginine (12.43 μg/l) and valine 11.99 (μg/l). An average of 15% of amino acids passed from cultivated fields are bound to humus substances. The highest concentration of these substances was measured for glycine (3.48 μg/l), arginine (2.39 μg/l) and cysteic acids (2.33 μg/l) (Table 4.4).

Table 4.4 Content of amino acids in dry mass of the groundwater flows to the shelterbelt and in humic substances isolated from the groundwater and in humic acids from soil

Amino Acids	*Total Amino Acids Flow to the Shelterbelt in the Groundwater in μg/l*	*Bound Amino Acids to Humic Substances in the Groundwater in μg/l*		*Amino Acids in Humic Acids in mg/kg of Humic Acids*	
		The ground-water flows to the shelterbelt	*The ground-water 16.5 m of the edge of the shelterbelt*	*In border between the field and the shelterbelt*	*In soil 16.5 m of the edge of the shelterbelt*
Acidic	15.6674	5.8840	3.0100	0.7061	1.0484
Neutral	68.0608	8.7664	6.4659	1.7968	1.5271
Basic	49.7575	4.5039	2.2377	0.6165	0.6883
Total Amount	**133.4857**	**19.1543**	**11.7136**	**3.1194**	**3.2588**

The following amino acids, however, were not determined for α-aminoadipic acid and citrulline. During outflows of groundwater across the 17 m wide afforestation, the content of bound amino acids to humic substances decreased by 39%. Acidic, neutral and basic groups of amino acids decreased the concentrations passed from afforestation. The highest decrease of basic and acid amino acids was observed to 50% and 44%, respectively. During flows of the groundwater across the shelterbelt the concentration of neutral amino acids decreased in the smallest degree (26%). This may indicate simultaneously a decrease of amino acid concentrations through afforestation and the supply of these substances by root exudates (Vancura 1967, Trojanowski 1973, Smith 1976).

CONCLUSIONS

The groundwater passing from the enjoining cultivated fields to the shelterbelt contains a high content of organic chemical compounds of well-known and unknown structures.

The 17 m-wide shelterbelt significantly decreases the concentrations of the following substances in the groundwater: total organic substances, total nitrogen and free and bound amino acids in humic substances.

Moreover, the groundwater passing from the shelterbelt decreases in the soil urease activity as well as pH and amino acid content in HAs.

REFERENCES

Allen, S.E. 1974. Organic constituents. pp. 237–301. *In*: Allen, S.E. (ed.), Chemical Analysis of Ecological Materials. Blackwell Scientific Publications, Oxford.

Bailey, R.A., Clarke, H.M., Ferris J.P., Krause, S. and Strong, R.L. 1978. Chemistry in aqueous media. pp. 318–360. *In*: Bailey, R.A., Clarke, H.M., Ferris, J.P., Krause, S. and Strong, R.L. (eds), Chemistry of the Environment, Academic Press Inc., New York.

Boratyński, K. 1981. Agricultural Chemistry. PWRiL, Warszawa, p. 407 (in Polish).

Buckman, H. and Brady, N.N. 1971. The Nature and Properties of Soil (In Polish). PWRiL, Warszawa, pp. 328–333.

Burm, J.R., Krawczyk, D.F. and Harlow, G.L. 1968. Chemical and physical comparison of combined and separate sewer discharge. Journal Water Pollution Control Federation. 40: 112–126.

Clark, M.D. and Gilmour, J.T. 1983. The effect of temperature on decomposition at optimum and saturated soil water contents. Soil Science Society America Journal. 47: 927–927.

Claudius, G. and Merhotka, R. 1973. Root exudates from lentil /Lens culinaris Madic/ seedlings in relation to wilt disease. Plant and Soil. 39: 315–320.

Connors, K.A. 1990. Chemical Kinetics. The Study of Reaction Rates in Solution, VCH Publishers Inc., pp. 17–309.

Correll, D.L. 1997. Buffer zones and water quality protection: General principles. pp. 7–20. *In*: Haycock, N.E., Burt, T.P., Goulding, W.T. and Pinay, G. (eds), Buffer Zones: Their Processes and Potential in Water Protection. Quest Environmental, Harpenden, U.K.

Crawford, N.H. and Donigian, A.S.Jr. 1973. Pesticide Transport and Runoff Model for Agricultural Lands. Office of Research and Development, U.S. Environmental Protection.

Donigian, A.S.Jr. and Crawford, N.H. 1976. Modeling Nonpoint Pollution from the Land Surface. Office of Research and Development, U.S. Environmental Protection Agency, EPA-600/3-76-083.

Donigian, A.S.Jr., Beyerlein, D.C., Davis, H.H.Jr. and Crawford, N.H. 1977. Agricultural Runoff Management (ARM) Model–Version II: Testing and Refinement, U.S. Environmental Protection Agency, EPA-600/3-77-098.

Eggen, T. and Majcherczyk, A. 2006. Effects of zero-valent iron (Fe°) and temperature on the transformation of DDT and its metabolites in lake sediment. Chemosphere. 62(7): 1116–1125.

Enfield, C.G., Phan, T. and Walters, D.M. 1981. Kinetic model for phosphate transport and transformation in calcareous soils. II. Lab and field transport. Soil Science Society America Journal. 45: 1064–1070.

Fitch, A., and Stevenson, F.J. 1984. Comparison of models for determining stability constants of metal complex with humic substances. Soil Science of America Journal. 48: 1044–1049.

Frost, A.A. and Pearson, R.G. 1961. Kinetics and Mechanisms. A Study of Homogeneous Chemical Reactions. John Wiley and Sons Inc., pp. 8–405.

Goring, C.A.I. 1967. Physical aspects of soil in relation to the action of soil fungicides. Annual Review Phytopathology. 5: 285–318.

Goring, H.K. and Lindahl, I.L. 1975. Growth and metabolism of sheep fed rations containing alfalfa hay of dehydrated alfalfa. Journal Dairy Science. 58: 759–765.

Hallberg, G.R. and Keeney, D.R. 1993. Nitrate. pp. 297–322. *In*: Alley, W.M. (ed.), Regional Ground-Water Quality. NewYork, Van Nostrand Rheinhold.

Harworth, R.D. 1971. The chemical nature of humic acid. Soil Science. 106: 188–192.

Hayes, M.H.B. and Swift, R.S. 1978. The chemistry of soil organic colloids. pp. 179–320. *In*: Greenland, D.J. and Hayes, M.H.B. (eds), The Chemistry of Soil Constituents. John Wiley and Sons, Chichester.

Helling, C.S., Kearney, P.C. and Alexander, M. 1971. Behavior of pesticides in soils. pp. 147–240. *In*: Brady, C.N. (ed.), Advances in Agronomy, Vol. 23. Academic Press, New York.

Hermanowicz, W., Dożańska, W., Dojlido, J. and Koziorowski, B. 1976. Physic-Chemical Examination of Water and Sludge. Warszawa, Arkady, p. 847 (in Polish).

Khaleel, R., Foster, G.R., Reddy, K.R., Overcash, M.R. and Westerman, P.W. 1979a. A non-point source model for land areas receiving animal waste: III. A conceptual model for sediment and manure transport. Trans. ASAE. 22: 1353–1361.

Khaleel, R., Foster, G.R., Reddy, K.R., Overcash, M.R. and Westerman, P.W. 1979b. A non-point source model for land areas receiving animal waste: IV. Model inputs and verification for sediment and manure transport. Trans. ASAE. 22: 1362–1368.

Khaleel, R., Reddy, K.R. and Overcash, M.R. 1980. Transport of potential pollutants in runoff water from areas receiving animal wastes: A review. Water Research. 14: 421–436.

Khaleel, R., Reddy, K.R. and Overcash, M.R. 1981. Changes in soil physical properties due to organic waste applications: A review. Journal Environmental Quality. 10: 133–141.

Kwak, J.C., Ayub, A.L. and Shepard, J.D. 1986. The role of colloid science in peat dewatering: Principles and dewatering studies. pp. 95–118. *In*: Fuchsman, C.H. (ed.), Peat and Water. Aspects of Water Retention and Dewatering in Peat. Elsevier Applied Science Publishers, London.

Lasaga, A.C., Berner, R.A., Fisher, G.W., Anderson, D.E. and Kirkpatrick, J.R. 1981. Kinetics of geochemical processes. pp. 1–68. *In*: Lasaga, A.C. and Kirkpatrick, J.R. (eds), Review in Mineralogy. Mineralogy Society of America.

Lityński, T., Jurkowska, H. and Gorlach, E. 1976. The Analysis in Chemistry and in Agriculture. PWN, Warszaw, p. 330 (In Polish).

Loehr, R.C. 1984. Pollution Control for Agriculture. Academic Press Inc., Orlando, pp. 1–455.

Luckner, M. 1990. Secondary Metabolism in Microorganisms, Plants and Animals, 3rd Ed. Springer-Verlag, Berlin, p. 576.

Overcash, M.R. and Davidson, J.M. 1980. Environmental Impact of Nonpoint Source Pollution. Ann Arbor Science Publishers, Ann Arbor, MI. p. 315.

Overcash, M.R., Khaleel, R. and Reddy, K.R. 1980. Nonpoint source model: Watershed inputs from land areas receiving animal wastes. Final Report EPA-R805011-01.

Overcash, M.R., Reddy, K.R. and Khaleel, R. 1981. Chemical and biological processes influencing the modeling of nonpoint source water quality from land areas receiving wastes. pp. 1–11. *In*: Overcash, M.R., Reddy, K.R. and Khaleel, R. (eds), Hydrologic Transport Modeling Water Resources Publ. Littleton, CO.

Pearson, G. 1996. Landscape diversity: A chance for the rural community to achieve a sustainable future. Report of the 2nd Pan-European Seminar on Rural Landscapes organized by th Council of Europe, the Ministry of Environment Protection, Natural Resources and Forestry of Poland and the Research Centre for Agricultural and Forest Environment of the Polish Academy of Sciences. Poznan (Poland) 25–30 September 1995. Council of Europe Publishing, Environmental Encounters, 26: 1–35.

Peterson, P.J. 1983. Adaptation of toxic metals. pp. 51–69. *In*: Robb, D.A. (ed.), Metals and Micronutrients, Uptake and Utilization by Plants. Pierpoint. Academic Press, London.

Reddy, K.R., Overcash, M.R., Khaleel, R. and Westerman, P.W. 1980a. Phosphorus adsorption-desorption characteristics of two soils utilized for disposal of animal wastes. Journal Environmental Quality. 9: 86–92.

Reddy, K.R., Khaleel, R. and Overcash, M.R. 1980b. Carbon transformations in the land areas receiving organic wastes in relation to nonpoint source pollution: A conceptual model. Journal Environmental Quality. 9(3): 434–442.

Reddy, K.R. 1982. Mineralization of nitrogen in organic soils. Soil Science Society America Journal. 46: 561–566.

Reddy, K.R. and Rao, P.S.C. 1983. Nitrogen and phosphorus fluxes from a flooded organic soil. Soil Science. 136: 300–307.

Reddy, K.R. 1983. Nitrogen and phosphorus cycling in shallow reservoirs used for agricultural drainage water treatment. pp. 407–426. *In*: Lowrance, R.R., Todd, R.L., Asmussen, L.E. and Leonard, R.A. (eds), Nutrient Cycling in Agricultural Ecosystems. Spl. Publ. 23. The Univ. Georgia Expt. Sta., Athens, GA.

Ryszkowski, L. 1989. Control of energy and matter fluxes in agricultural landscape. Agriculture, Ecosystems and Environment. 27: 107–118.

Ryszkowski, L., Kędziora, A., Bartoszewicz, A., Szajdak, L. and Życzyńska-Bałoniak, I. 1998. Influence of agricultural landscape diversity on water and nitrogen cycling. Farina, A., Kennedy, J. and Bossiù, V. (eds), Book of Abstract. Intecol. VII, International Congress of Ecology. Florence 19–25 July, Italy, p. 365.

Ryszkowski, L., Szajdak, L., Bartoszewicz, A. and Życzyńska-Bałoniak, I. 2002. Control of diffuse pollution by mid-field shelterbelts and meadow strips. pp. 111–143. *In*: Ryszkowski, L. (ed.), Landscape Ecology in Agroecosystems Management. CRS Press, Boca Raton.

Schnitzer, M. and Khan, S.U. 1978. Soil Organic Matter. Elsevier Scientific Publishing Company, Amsterdam, pp. 261–262.

Sewell, J.I. and Alphin, J.M. 1975. Effects of agricultural land uses on runoff quality. Tennessee, Agricultural Experiment Station, Knoxville Bulletin. 548: 44–47.

Sharpley, A.N. and Smith, S.J. 1992. Prediction of bioavailable phosphorus loss in agricultural runoff. Journal Enironmwntal Quality. 21: 32–37.

Smith, W.H. 1976. Character and significance of forest tree root exudates. Ecology. 57: 324–331.

Stevenson, F.J. 1982. Humus Chemistry, Genesis, Composition, Reaction. J. Wiley and Sons, New York, pp. 285–320.

Szajdak, L. 1996. Impact of crop rotation and phenological periods on rhodanese activity and sulphur amino-acids concentrations in brown soils under continuous rye cropping and crop rotation. Evironmental International. 22(5): 563–569.

Szajdak, L. and Österberg, R. 1996. Amino acids present in humic acids from soils under different cultivations. Environment International. 22(5): 331–334.

Szajdak, L. and Matuszewska, T. 2000. Reaction of woods changes of nitrogen in two kinds of soil. Polish Journal of Soil Science. XXXIII/1: 9–17.

Szajdak, L. and Meysner, T. 2002a. Influence of the shelterbelt on the containment of amino acids bound in humic acids in various kinds of soils. Polish Journal Soil Science. 35(1): 47–57.

Szajdak, L., Maryganova, V. and Meysner, T. 2002a. The function of shelterbelt as biogeochemical barrier in agricultural landscape. Acta Agrophysics. 67: 263–273.

Szajdak, L., Maryganova, V., Meysner, T. and Tychinskaja, L. 2002b. Effect of shelterbelt on two kinds of soils on the transformation of organic matter. Environmental International. 28: 383–392.

Szajdak, L., Życzyńska-Bałoniak, I. and Jaskulska, R. 2003a. Impact of afforestation on the limitation of the spread of the pollutions in ground water and in soils. Polish Journal Environmental Studies. 12(4): 453–459.

Szajdak, L., Jezierski, A. and Cabrera, M.L. 2003b. Impact of conventional and no-tillage management on soils amino acids, stable and transient radicals and properties of humic and fulvic acids. Organic Geochemistry. 34: 693–70.

Szajdak, L., Życzyńska–Bałoniak, I., Jaskulska, R. and Szczepański, M. 2009. Changes of the concentrations of dissolved chemical compounds migrating into ground water through biogeochemical barriers in an agricultural landscape. Oceanological and Hydrobiological Studies. International Journal Oceanography Hydrobiology. 38(4): 109–116.

Szpakowska, B. and Życzyńska–Bałoniak, I. 1994a. Groundwater movement of mineral elements in an agricultural area of southwestern Poland. pp. 95–104. *In*: Ryszkowski, L., and Bałazy, S. (eds), Functional Appraisal of Agricultural Landscape in Europe. Zakład Badań Środowiska Rolnicznego i Leśnego PAN, Poznań.

Szpakowska, B. and Życzyńska-Bałoniak, I. 1994b. The role of biogeochemical barriers in water migration of humic substances. Polish Journal of Environmental Studies. 3: 35–41.

Trojanowski, J. 1973. The Conversion of Organic Substances in Soil. PWRiL, Warszawa, p. 331 (in Polish).

Vancura, V. 1967. Root exudates of plant. III. Effects of temperature and "cold shock" on the exudation of varius compounds from seeds and seedlings of maize and cucumber. Plant and Soil. 27: 319–328.

Walker, A. 1987. Evaluation of a simulation model of herbicide movement and persistence in soil. Weed Research. 27: 143–152.

Walker, A. and Hollis, J.M. 1994. Prediction of pesticide mobility in soils and their potential to contaminate surface and groundwater. pp. 211–224. *In*: Hewitt, H.G. (ed.), Comparing Glasshouse and Field Pesticide Performance. BCPC Monograph no. 59, British Crop Protection Council, Farnham, UK.

Walker, A., Calvet, R., Del Re, A.A.M., Pestemer, W. and Hollis, J.M. 1995. Evaluation and Improvement of Mathematical Models of Pesticide Mobility in Soils and Assessment of their Potential to Predict Contamination of Water Systems. Mitteilungen aus der Biologischen Bundesanstalt für Land- und Forstwirtschaft, Berlin-Dahlem, 307: Blackwell Wissenschafts-Verlag, Berlin, pp. 32–47.

Walker, A., Welch, S.J., Melacini, A. and Moon, Y.-H. 1996. Evaluation of three pesticide leaching models with experimental data for alachlor, atrazine and metribuzin. Weed Research. 36: 37–47.

Życzyńska-Bałoniak, I, Szpakowska, B., Ryszkowski, L. and Pempkowiak, J. 1993. Role of meadow strips for migration of dissolved organic compounds and have metals with groundwater. Hydrobiologia. 251: 249–256.

Życzyńska-Bałoniak, I., Ryszkowski, L. and Waack, A. 1998. Transport wodą gruntową rozpuszczonych związków chemicznych poprzez zadrzewienia. Zeszyty Problemowe Postępów Nauk Rolniczych. 460: 167–176.

Życzyńska-Bałoniak, I., Szajdak, L. and Jaskulska, R. 1996. Organic compounds in the water and their function in agriculture landscape. pp. 133–142. *In*: Ryszkowski, L. and Bałazy, S. (eds), Cycle of Water and Biochemical Barriers in Agricultural Landscape. Zakład Badań Środowiska Rolniczego i Leśnego PAN, Poznań. (in Polish).

Życzyńska-Bałoniak, I. and Szpakowska, B. 1989. Organic compounds dissolved in water bodies situated in an agricultural landscape and their role for matter migration. Archiv Hydrobiology Beihefte *Ergebnisse* Limnologie. 33: 315–322.

Chapter 5

Dynamics of Fertilizer Application and Current State of the Nitrogen Balance in the Soils of the Republic of Moldova

Olesea Cojocaru[1*] and Lech Wojciech Szajdak[2]

[1]Agrarian State University of Moldova, Chisina, Republic of Moldova.

[2]Institute for Agricultural and Forest Environment, Poznań, Poland.

INTRODUCTION

Soil presents the environment where the most radical transformations of organogenic elements (carbon, oxygen, hydrogen and nitrogen) occur. Nitrogen transformation processes directly involve soil biota and green plants, which develop according to pedoclimatic conditions. When pedoclimatic conditions are favorable, the intensity of the processes of nitrogen transformation in the soil is optimal and vice versa. The arable soil of Moldova contains total nitrogen between 0.05% and 0.23% by weight (Table 5.1). The general state of the nitrogen circuit is reflected by its total content in different layers of the soil profile. Total nitrogen in the soil encompasses all accumulation and consumption patterns and characterizes its potential fertility.

Total nitrogen is a component of soil humus. It is established that in the composition of humus, carbon contains 50%, oxygen is 39%, hydrogen is 5% and nitrogen is 5% of the dry mass. Other macro and micronutrients are 1% (Andrieş et al. 2001, Кордуняну et al. 1997, Понамарева et al. 1980). Although the nitrogen

*For Correspondence: E-mail: o.cojocaru@uasm.md

content of the humus is on average 5%, it plays an important role in all processes of soil humus transformation. The total nitrogen content and carbon ratio (C:N) of the organic mass characterize its effective fertility. The soil type mold report is 9–14 units (Andrieş et al. 2001, Banaru 2002, Кордуняну et al. 1997, Понамарева et al. 1980, Щербаков 1968).

Nitrogen is used by plants throughout the growing season in large quantities. For the formation of 5.0 t/ha of autumn, wheat plants extract from the soil is 150 kg/ha of nitrogen. Moldova's soil produces annually, after biological processes, 75 kg/ha of mineral nitrogen (N_{min}), which is available to plants. The eroded soils, which occupy about 40% of the agricultural land, are characterized by a lower nitrification capacity and produce only 30–60 kg/ha per year. Nitrogen deficiency in plant nutrition leads to yellowing of the leaves, slowing or stopping their growth. The main causes of nitrogen deficiency in plant nutrition are soil properties that inhibit the activity of microorganisms (low content of organic matter, secondary compaction, degradation, etc.), the absence of cropping of leguminous crops, atmospheric nitrogen fixators, the non-application of organic fertilizers or their introduction in small doses, inadequate soil work and unreasonable use of plant debris.

In the past 15–20 years, agricultural crops are not being respected; organic fertilizers are applied at low doses and the volume of mineral fertilizers has been considerably reduced and is 10–50 kg/ha. The production capacity of the soil has decreased by 25–35%. As a result, the autumn wheat harvest in a multiannual cycle is only 2.5–2.8 t/ha. The purpose of the investigations is to evaluate the current state of the Moldovan soils, forms and reserves of nitrogen and to develop measures for the optimization of the mineral nutrition of the crop plants (Burlacu 2000).

The nitrogen circuit includes nitrogen excess and deficiency. The circuit of nitrogen between and in the essential components of ecosystems—soil, plants, water and air resources—hold physicochemical, biochemical and microbiological processes with inputs of N (a) substances and, in contrast, another sum of natural phenomena and/or anthropic ones leading to the diminution of resources with N (b).

a) N-input processes in the soil-plant circuit: Here are the mineralization of humus and organic nitrogen with the nitrification and ammonification sequences, the symbiotic and non-imbibition of nitrogen, the net N inputs with the industrial and organic fertilizers and the intake of N with precipitation.
b) Processes with reduction and/or removal of nitrogen from the soil-plant circuit: Here is the consumption of N for the formation of the crops (according to Cs and Cg), the microbial immobilization of the mineral and organic nitrogen, $N–NO_3$ losses by percolation in the soil, N by ammonium fixation, denitrification and erosion losses. The separate quantification of the positive and negative processes in the nitrogen regime allows a balance of this essential element at ground level and practical technological decisions to intervene or to rationalize the balance sheets in the production of the quantitative and qualitative vegetal production and the protection of the resources in the environment.

Excess of N- is interpreted in agricultural practice as a result of the application of large and excessive fertilizers over the ability to assimilate crops and horticultural crops. This excess of N-mineral (NO_3^- + NH_4^+) occurs at doses of N that exceed 200–250 kg. N/ha applied at one time, unfractionated with a poor supply of P, K and Ca + Mg. The immediate effect of this surplus of $N_{mineral}$ is the excessive and phytotoxic accumulation in the green organs of the plants (mostly leaves) of the mineral forms of N (NO_3^- first) with the symptom of "nitric toxicity", subsequently with the compromise of the culture and in the soil, and in the percolative water stream, part of the surplus of N (mostly in the form of NO_3^-) contaminates the soil and groundwater profile. Soil acidity favors the negative effects of excess N this being itself accentuated by excessive doses of N (Andrieş et al. 2001, Кордуняну et al. 1997, Понамарева et al. 1980).

Prevention of excess N can be achieved as follows: optimal doses of N applied fractionally as mineral fertilizers—taking into account the effective input of the soil in the ions of this element—correct reporting to the content and intake of P, K, Ca, Mg, S and microelements—Mo, correcting or protecting the reaction with calcareous modifications between fertilizers, diversified crops consuming N in crop production genotypes, etc.

Excess N-crops contain >3.5–4% N-total with more than 1,500–2,000 ppm; N-mineral in various plant organs with <0.15% P_2O_5; <1–1.50% K_2O; 1.0% CaO in dry soil. The soil contains on the agricultural profile (0–100 cm) over 250–300 kg/ha of $N_{mineral}$, part of which is outside the edaphic volume of plants.

Insufficient N naturally belongs to poor humus (<1–1.5% H) and below 20–30 ppm N-mineral. They are native soils, sandy, young alluvium but also eroded soils (erodosols). In plants the N_t content <1–1.5% in dry soil with a yellowing symptom of the phenomenon is determined. Increases are weak, poor twinning of cereals, leaves and small fruit, quantitatively and qualitatively compromise the culture. N-insufficiency can also be determined by uncontrolled agricultural technology (without fertilizers). Prevention of N-deficiencies and deficiencies is accomplished by fertilization with N fertilizers but well correlated with the other elements (fertilizers) that boost the chemistry and effect (P, K, S, Ca and Mg).

Fertilization should take into account the positive, multi-annual evolution of the complex N-system in the soil-plant system. Some crops that respond positively to organomineral fertilization (trees, vines and greenhouse and field vegetables) can also promote some organic fertilizer resources or the incorporation of plant debris from plants positively supported by minimal N or NPK interventions to encourage humus synthesis as one side of the long-term improvement of the N-system and the supply of soils and crops.

MATERIAL AND METHOD

The research was conducted in the years 1985–2015 in the long-term field experiments of the Institute of Pedology, Agrochemistry and Soil Protection "Nicolae Dimo", based on gray soil, leachate, ordinary and carbonate chernozems. The agrochemical characterization of soils and some results obtained in field experiments with the application of fertilizers in crops are presented in a series of monographs (Andrieş 2007, 2011, Donos 2008, Lupaşcu 2004).

Table 5.1 Average nitrogen content in the soils of the Republic of Moldova, % of the dry mass

No	*The Name of the Soil*	***Soil Layer*, cm (%)**				**0–50 cm, t/ha**
		0–20	20–30	30–40	40–50	
1	Brown soil with a full profile	0.26	0.08	0.05	0.04	8.0
2	Gray soil with a full profile	0.14	0.13	0.11	0.06	7.0
3	Gray soil eroded weak	0.11	0.10	0.09	0.05	5.5
4	Gray soil eroded moderate	0.08	0.07	0.06	0.04	4.0
5	Gray soil eroded strong	0.05	0.04	0.03	0.02	2.3
6	Argiloiluvial chernozem with a full profile	0.21	0.19	0.18	0.15	11.4
7	Leached chernozem with a full profile	0.23	0.22	0.20	0.17	12.7
8	Leached chernozem eroded weak	0.21	0.20	0.18	0.15	11.3
9	Leached chernozem eroded moderate	0.18	0.15	0.12	0.10	8.8
10	Leached chernozem eroded strong	0.15	0.13	0.10	0.08	7.4
11	Typical chernozem with a full profile	0.24	0.23	0.21	0.18	13.4
12	Typical chernozem eroded weak	0.22	0.19	0.17	0.14	11.4
13	Typical chernozem eroded moderate	0.19	0.17	0.14	0.12	9.8
14	Typical chernozem eroded strong	0.16	0.13	0.11	0.09	7.8
15	Ordinary chernozem with a full profile	0.21	0.20	0.19	0.17	11.9
16	Ordinary chernozem eroded weak	0.19	0.18	0.16	0.14	10.4
17	Ordinary chernozem eroded moderate	0.17	0.15	0.13	0.12	8.9
18	Ordinary chernozem eroded strong	0.15	0.13	0.12	0.10	7.8
19	Carbonated chernozem with a full profile	0.21	0.20	0.19	0.17	11.9
20	Carbonated chernozem eroded weak	0.19	0.18	0.16	0.17	10.4
21	Carbonated chernozem eroded moderate	0.17	0.16	0.14	0.12	9.3
22	Carbonated chernozem eroded strong	0.15	0.14	0.13	0.11	8.5

Note: Samples for eroded soil have been collected from gray soil areas and not eroded chernozem subtypes.

Source: Data from Donos Alexei, "Acumularea şi transformarea azotului în sol". 2008: 54–55.

During the period 1989–2010, early spring, the agrochemical mapping of the soil was carried out in different pedoclimatic areas on 1,500–4,500 ha. Mineral nitrogen reserves were determined in the 0–100 cm layer and the moisture content 0–160 cm thick, the mobile phosphorus content and the exchangeable potassium and the phenophase of the grain development. The data of the statistical yearbooks of the Republic of Moldova on the application of organic and mineral fertilizers in agriculture, the area occupied by the plants and their harvest were analyzed. The experimental results were processed by different statistical methods. Regarding the total nitrogen accumulated in the soils of the Republic of Moldova, it was calculated that 7–9 t/ha of nitrogen was contained in the arable layer of the leachate and typical chernozems and in the layer thickness of 1 m the total nitrogen content is 18.4–21.5 t/ha. Data obtained from multiple field investigations, which are systematized in Table 5.1, demonstrate that in the most biogenic soil layer of

0–50 cm the nitrogen content in different soil types and subtypes is different (Andrieş 2007, 2011, 2017, Andrieş et al. 2000, 2001a, 2001b, Крупеников 1967, 2008, Синкевич 1989).

The use of sulfuric acid of various concentrations allows determining the structure of organic nitrogen compounds substance (Шконде 1965). Using this method, we investigated the structure of the non-hydrolyzable organic nitrogen fraction, which is hardly hydrolyzable and easily hydrolyzable. These forms of organic nitrogen make up 85–97% of the total (Table 5.2).

RESULTS AND DISCUSSIONS

Nitrogen from soil has accumulated over hundreds and thousands of years and is closely related to organic matter. The coefficient of correlation between nitrogen and carbon content is 0.82–0.98 (Думитрашко 1987). The reduction of molecular nitrogen in the atmosphere occurs after its fixation by symbiotic bacteria with inclusion in the microbial mass, later in the biomass of plants and animals. Organic substances accumulated in the soil are subject to decomposition and mineralization processes with the formation of humus and mineral nitrogen forms (Tan 2000).

Content and Reserves of Nitrogen in the Soil

The soils of Moldova are characterized by relatively high nitrogen content. It has been established that about 1 billion tons of humus and 50 million tons of nitrogen have accumulated in the soil. As shown in the generalized data (n = 1,527), the total nitrogen content in the chernozem is 0.21–0.24%. Nitrogen reserves in the 0–20 cm chernozem layer are 5.5 t/ha. In forest soils (brown and ash) the total nitrogen content is less than in chernozems, and it is 3.9 t/ha (Думитрашко 1987). Nitrogen in soil is contained in organic and inorganic form. The organic form of nitrogen in the soil constitutes 96–97% of the total (Donos 2008).

Organic nitrogen consists of the following fractions: non-hydrolyzed, heavily hydrolyzed and slightly hydrolyzed. The main organic N content in the soil (about 70–80% of the total) is represented by non-hydrolyzed organic compounds in concentrated acid solutions such as humines, melanins and bitumens. Non-hydrolyzed nitrogen plays a prime role in forming and maintaining the soil nitrogen structure. The hardly hydrolyzed fraction is 10–15% of organic nitrogen and represents the N-protected reserve for plant nutrition. This fraction is involved in completing and maintaining the nitrogen of the soil. The fractions of non-hydrolyzed and heavily hydrolyzed nitrogen are part of the stable organic matter (humus), estimated at 800 to 3,000 years (Rusu 2008). The fraction of the slightly hydrolyzed nitrogen fraction is 8–15%. This fraction includes both organic and inorganic compounds.

The slightly hydrolyzed N-fraction is the main source of supplementation of reserves of organic nitrogen in the soil and mineral nitrogen ($N_{M\ in}$), which is accessible to plants. The nitrogen of this fraction is part of the labile organic matter, which in the soils of Moldova is estimated at 10–15% of the total

(Andrieş 2017, Андрианов et al. 1989). The mineral forms of nitrogen in the soil are fixed ammonium, 2–3%, ammonium exchangeable, nitrates, nitrites and nitrogen in the gaseous form. Changing ammonium ions are adsorbed to the surface of colloidal micelles, and those in the form of the anchor are fixed in the spaces between the sheets of clay minerals. The exchangeable ammonium form of the soil has a short existence as it is absorbed by plants or subjected to biochemical processes and transformed into nitrates. The amount of mineral nitrogen constitutes up to 1–2% of the total (Camberado 2012, Donos 2008). In the years 1989–2010, the amount of $N_{mineral}$ from the soil under the conditions of production was monitored for the cultivation of autumn wheat.

According to the data in Table 5.1, the largest amounts of total nitrogen are contained in the typical chernozem 13–14 t/ha, followed by the ordinary, leachate, carbonate and argiloiluvial chernozems. In gray soils, the total nitrogen reserve is 7–8 t/ha, which at 50–75% is concentrated in the upper layer of 0–20 cm of arable soil. In eroded soils, the total nitrogen content decreases by 20–70% compared to the full profile soil types. For example, eroded weak ordinary and carbonate chernozems contain 20–30% less total nitrogen than full-soil soils. In moderately eroded soils, this amount of nitrogen decreases by 40–50%, and in strong eroded soils by 51–70% compared to not eroded soils.

Thus, in weakly eroded soils, total nitrogen reserves range from 5.5 t/ha in the 0–50 cm layer of gray soil to 11.3 t/ha in the typical chernozem. The strongly eroded soils contain from 2.3 t/ha in gray soil to 7.8 t/ha in the strongly eroded typical chernozem. After the total nitrogen reserves in the arable layer of 0–20 cm, the richest is the typical brown soil that is 5.9 t/ha. In fully chernozemed soils with a full profile, in this bioactive layer, the total nitrogen reserve ranges from 4.8 to 5.5 t/ha.

Even in eroded soils most of the nitrogen is contained in the arable layer. This is due to the fact that the main mass of vegetal remains is formed and remains in it (Кордуняну et al. 1997). Undoubtedly, total nitrogen characterizes the potential fertility of the soil and, along with the humus, is one of the main indicators.

The non-hydrolyzable organic nitrogen in the soils of Moldova constitutes 65–68% of the total and is represented by stable compounds, which do not decompose in the soil hydrolysis process with 5N sulfuric acid solution (Table 5.2).

Studies have also shown that the systematic application of mineral fertilizers ($N_{120}P_{60}K_{60}$) increases the non-hydrolyzable nitrogen content by 3–4% compared to unfertilized variants. We conclude that in the soil, up to the collection of samples, we applied in the gray soil 1,200 kg/ha N with ammonium nitrate, in the leachate chernozem 1,320 kg/ha N and in the ordinary chernozem 480 kg/ha N.

The calculation of the non-hydrolyzable organic nitrogen stock found a higher quantity compared to the control by 281 kg/ha, 387 and 118 kg/ha respectively, on the fertilized variants, which represented 23.4–29.3%; 24.6% of the total amount of fertilizer applied to gray soil, leachate and ordinary chernozem.

An increase in the content of organic nitrogen in the soil application of nitrogen fertilizers have found a number of researchers (Андриеш 1993, Донос 1977a, 1977b, 1978a, 1978b, 1978c, 1979, Донос and Кордуняну 1978, Смирнов 1977, 1979, Щербаков 1968, Шконде 1965, Загорча 1990).

Table 5.2 Content of nitrogen forms in the 40 cm layer of arable soils of the Republic of Moldova

Forms of Nitrogen	*% of the dry soil mass*		*in t/ha*		*in % of total nitrogen*	
	Average Deviation Limits	*Average*	*Average Deviation Limits*	*Average*	*Average Deviation Limits*	*Average*
Mineral nitrogen (NH_4+NO_3)	0.001–0.012	0.009	0.03–0.15	0.08	0.4–1.7	0.8
Ammoniacal nitrogen fixed to minerals	0.005–0.018	0.012	0.30–0.60	0.50	3.0–6.0	4.5
Easily hydrolyzable organic nitrogen	0.008–0.03	0.023	0.70–1.60	1.10	8.0–13.0	11.5
Hardly hydrolyzable organic nitrogen	0.016–0.04	0.026	1.00–1.70	1.30	12.0–16.0	15.6
Non-hydrolyzable organic nitrogen	0.10–0.13	0.11	5.00–7.50	6.0	65.0–68.0	67.6
Total nitrogen	0.13–0.28	0.18	7.00–12.5	9.0	100.0	100.0

Source: Data from Donos Alexei, “Acumularea şi transformarea azotului în sol”. 2008: 56–57.

For example, Смирнов (1979, p. 56), the experiences in the field, using 70–120 kg/ha of ammonium sulfate found that the fraction of organic nitrogen was fixed 32–52%, average of 44.1% of the applied dose. When using ammonium nitrate in doses of 70–120 kg/ha, the organic nitrogen fraction was set at 20–31%, the mean of 24.8% of the applied nitrogen dose.

At the same time, intensive and long-term use of the soil without the application of fertilizers leads to the quantitative reduction of the non-hydrolyzable organic nitrogen in the soil, especially from the arable layer. The non-hydrolyzable nitrogen content of the ordinary chernozem, used in the agricultural circuit, is 29–30% lower than in the unworked ordinary chernozem. On wooded land, the non-hydrolyzable organic nitrogen content is 27–31% higher than in the soil used in agriculture.

As determined, the amount of $N_{mineral}$ depends largely on the sources, forms and amount of nitrogen stored in the soil (Table 5.3). In the 1980s and 1990s, considerable amounts of fertilizers were incorporated into the soil: 5.0–6.5 t/ha of organic fertilizers and 50–60 kg/ha of mineral nitrogen. In the field crops, the share of perennial grasses was relatively high and constituted 8–12%. The area occupied by peas was 80–90 thousand hectares.

Vegetable remains were used as organic fertilizer. With such a flow of nitrogen in the soil the $N_{mineral}$ reserves were large (115 kg/ha), and the plants on 70% of the area surveyed were insured with mineral nitrogen. Only 10% of the investigated area was characterized by low $N_{mineral}$ content in the soil. In the years 1994–2015, another situation was created in providing plants with mineral nutrition, including nitrogen. During this period the nitrogen flow from the soil was substantially reduced. The share of leguminous crops, biological nitrogen sequestration, decreased fivefold.

Very small amounts of fertilizer (0.1–0.02 t/ha manure and 10–50 kg/ha NPK) were incorporated into the soil. Plant residues were not used everywhere as an organic fertilizer. As a result, $N_{mineral}$ reserves declined twice. On 70–75% of the area investigated, the amount of $N_{mineral}$ was low and the effectiveness of nitrogen fertilizers was high. Each kilogram of N was recovered with 10–35 kg of winter wheat. The technical-scientific production developed by the institute in the years 1994–2010 (winter wheat harvest forecast, $N_{mineral}$ reserves and soil humidity, N doses and fertilizer application times) was made operational by farmers and agricultural specialists for implementation.

Table 5.3 N–NO_3 content in the soil layer of 1 m soil in winter wheat, % of the area surveyed

Content of N–NO₃	*Reserve of N–NO₃, kg/ha*	*Years*				*The Need for Autumn Wheat in Nitrogen*
		1986 n*=77	**2003 n=51**	**2007 n=108**	**2010 n=17**	
Low	up to 60	9	70	75	70	High
Moderate	60–100	29	14	19	18	Moderate
Optimum	100–140	23	16	6	12	Low
Picked up	over 140	39	0	0	0	Missing
Average weighted of N–NO_3, in the layer of 1 m, kg/ha		115	56	46	51	

This action was carried out with the support of the Ministry of Agriculture and Food Industry. The information was disseminated through the media and agricultural seminars. At the current stage of agricultural development, the balance of organic matter and nitrogen in the soil is negative (Andrieş 2017, Lupaşcu 2004) and the plants are insufficiently assured with nutrients to obtain the expected yields of 4.0–5.0 t/ha of winter wheat.

Biochemical Transformations of Soil Nitrogen

The soil nitrogen regime depends on two biochemical processes: mineralization and immobilization. To obtain energy and nutrients soil microorganisms decompose organic matter through the mineralization process. As a result of this process in the soil, mineral forms of nitrogen are formed. The immobilization phenomenon is a process of passing a chemical element of an inorganic form into an organic form in microbial tissues or vegetal tissues, transforming it into inaccessibility for other microorganisms and plants (Lăcătuşu 2000).

The mineralization and immobilization processes are based on two factors: C:N ratio and time. The rate of decomposition of organic substances increases if the C:N ratio is less than 15. In such conditions, mineralization processes are predominant with the accumulation in the soil of mineral nitrogen forms. Organogenic materials with a C:N ratio greater than 30 is characterized by a low and very low decomposition rate. Under these conditions of decomposition of organic substances, the immobilization processes predominate. Thus, for the decomposition of straw cultures with C:N ratio of 50–100, maize stalks with the ratio C:N = 70 or wood sawdust with the ratio C:N = 200–500, wherein considerable amounts of nitrogen are required. For the decomposition of one tone of wheat autumn straw, it is recommended to apply 10 kg of nitrogen (Andrieş et al. 2001, Lupaşcu 2004, Programul complex de valorificare a terenurilor degradate şi sporirea fertilităţii solurilor. Partea II. 2004b).

Otherwise, the decomposition process is slow enough and the mineral nitrogen formed by biochemical processes in the soil will be immobilized by the microorganisms involved in the decomposition of straw spike cultures. At harvest of 4.0 t/ha of autumn wheat, 5.0 t/ha of straw is incorporated into the soil. For their decomposition, it is recommended to apply 50 kg/ha of nitrogen (140 kg/ha of ammonium nitrate). Nitrification is a process of biological oxidation of ammonium to nitrates by nitrifying bacteria in the soil. For the grouping of agricultural soils after the nitrification capacity, agrochemical investigations were conducted in all pedoclimatic zones of Moldova. The amount of organic matter in the soils studied ranges from 4.63 to 0.89%. For these soils, the nitrification capacity is 19.1 and 0.3 mg NO_3/100 g, respectively. The statistical processing of the experimental data allowed to show a close correlation between the nitrification capacity (Y, mg/100 g soil) and the soil organic matter content (X, %), expressed by the following equation: $Y = -6.71 + 7.99X - 0.67X_2$, r = 90; n = 75.

Based on the results obtained, the nitrification capacity of the soils was determined (Table 5.4). In soils with a low content of organic matter (less than 2.0%), the nitrification capacity is very low and is below 6.5 mg/100 g. Soils with an organic content of between 4% and 5% have an optimal nitrification capacity of 15–17 mg NO_3/100 g.

Table 5.4 Soil production capacity according to agrochemical indices

The Content of Organic Matter, %	***Nitrification Capacity,*** **mg/NO$_3$, 100 g**	***Nitric Nitrogen Content in the*** **0–100 cm, kg/ha**	***Forecast of Wheat Autumn Harvest,*** **t/ha**
< 2.0	< 6.5	< 48	< 1.7
2.1–3.0	7.0–11.1	50–72	1.7–2.4
3.1–4.0	11.5–14.6	74–96	2.4–3.2
4.1–5.0	14.9–17.1	98–120	3.2–4.0
> 5.1	> 17.3	> 122	> 4.1

Source: Data from Andrieş Serafim, "Optimizarea regimurilor nutritive ale solurilor şi productivitatea plantelor de cultură". 2007: 45–194.

Experimentally it was established (Andrieş 2007) that the increase of the organic matter content in soil by 1% ensures the formation of 24 kg/ha of mineral nitrogen. Soils with organic matter content of 2% produce annually 48 kg/ha of $N_{mineral}$, enough to form only 1.6 t/ha of winter wheat grains. Soils with organic matter content of 4.0%, such as typical and leached chernozems, produce 96 kg/ha of mineral nitrogen per year. This amount of $N_{mineral}$ ensures the formation of 3.2 t/ha of grain autumn wheat without the additional application of nitrogen fertilizers. Established agrochemical parameters can be used to assess nitrogen plant assurance and to determine soil production capacity.

Nitrogen Losses in the Soil

Nitrogen from the soil is lost due to leaching, denitrification, volatilization and erosion processes. Ligation is defined as a phenomenon of removal of soluble components, including nutrients, from the top of the soil to the bottom (Lăcătuşu 2000). Due to the high mobility of nitrates, they migrate with the downward flow, reaching pedophrematic waters. The nitrate leaching phenomenon has two aspects: considerable nitrogen losses from fertilizers and pollution of pedophreate water with nitrates.

Determination of nitrate content in long-term field experiments and production conditions up to 20 m depth (Andrieş 2007, Tate 2000) allowed the following conclusions to be drawn:

- The nitrate leaching under the influence of atmospheric precipitation occurs up to 9–10 m in the ordinary chernozem in the southern area and up to 10–11 m in the levigated chernozem in the central area of Moldova.
- Nitrogen losses from fertilizers through the leaching process are 8.8–9.3%.
- The nitrate leaching is more intense during the cold period of the year. During the warm period, the leaching of nitrogen with the downward flow takes place up to 20–60 cm, a layer of which nitric nitrogen is fully used by plants. The following measures (Andrieş 2007, 2011, Lupaşcu 2004, Programul complex de valorificare a terenurilor degradate şi sporirea fertilităţii solurilor. Partea II. 2004b) are indicated to minimize nitrogen leakage from the soil through leaching.
- Monitoring the mineral nitrogen reserves in agricultural soils.
- Determination of doses of nitrogen fertilizers based on the complex soil-plant diagnosis.
- Application of fertilizers with nitrogen, predominantly during the spring/summer period.
- The use of nitrogen fertilizers in several instances (in soil, foliage) during the maximum consumption period by crop plants.

By denitrification (the process of biological reduction of nitrates in the form of nitrogen gas, as molecular nitrogen or as oxides of nitrogen (Lăcătuşu 2000), nitrogen losses can constitute 5–20% of the amount of N applied as a fertilizer (Camberado 2012, Grahmann et al. 2013). The volatilization (volatilization process) of ammonia is one of the important ways of soil nitrogen loss (Lăcătuşu 2000). Ammoniacal nitrogen is lost more intensely on calcareous soils. Losses through volatilization can reach 20–50% on superficial application of anhydrous ammonia or urea on alkaline soils, wind and high temperatures. To avoid nitrogen leakage through volatilization it is recommended to incorporate nitrogen fertilizer into wet soil (Programul complex de valorificare a terenurilor degradate şi sporirea fertilităţii solurilor. Partea I. 2004a). Nitrogen losses through erosion are 27–29 kg/ha.

Larger leaks, implicitly nitrogen losses, occur in hoeing crops and in and vineyard plantations. For frequent crops, nitrogen losses through erosion are smaller and constitute 8–10 kg/ha. Some of the liquid spills from the slope lands penetrate the hydrographic network and pollute surface waters with nitrogen compounds. Combating soil erosion is achieved through the implementation of good agricultural practices with the minimization of solid spills up to the admissible limits of 5 t/ha (Nour 2004).

Efficacy of Nitrogen Fertilizers

The effect of fertilizer application depends largely on soil type and subtype, soil agrochemical properties, agrometeorological conditions, crop plant, etc. It has been established (Andrieş 2007) that the effectiveness of nitrogen fertilizers is closely related to the mineral nitrogen reserves in the soil. The increase in the autumn wheat harvest at the application of the N_{60} dose on the background of PK or NPK ranged from 67% to –12%. Between the reaction of plants at N_{60} (Y, %) and the quantity of $N–NO_3$ in the layer of 1 m in spring at the beginning of the vegetation period (X, kg/ha) a close correlation, expressed by the following equation: Y = (1,181: X) – 4.2; r = 0.71. Based on the experimental data, the $N_{mineral}$ stock gradient was calculated in the agricultural soils and the need for additional nitrogen nutrition of the winter wheat (Table 5.5) was determined.

On soils with very low mineral content (less than 35 kg/ha), the increase in harvest is high and exceeds 1.5 t/ha of winter wheat grains. Each kilogram of nitrogen is recovered with over 20–25 kg of grain. On soils with high $N_{mineral}$ availability of plants (over 140 kg/ha), autumn wheat does not react to supplemental nitrogen nutrition. Plants are secured with nitrogen in an optimal amount to form 4.5–5.0 t/ha of winter wheat when the mineral nitrogen reserves in the 1 m layer of soil form 100–140 kg/ha.

Table 5.5 Need for autumn wheat in nitrogen depending on the $N–NO_3$ reserves in the soil

The content of $N–NO_3$ in the Layer 0–100 cm, kg/ha		***Increase in Harvest at N_{60}***		***The Need for Autumn Wheat in Nitrogen***
		kg/ha	**%**	
Very low	under the 35	over 1,500	over 40	Very big
Low	36–60	1,000–1,500	20–40	Big
Moderate	61–100	300–1,000	5–20	Average
Optimum	101–140	0–300	0–5	Low
High	over 140	0	0	Missing

Source: Data from Andrieş Serafim, "Agrochimia elementelor nutritive". 2011: 27–73.

To confirm the above, we present the results of the field experiments conducted in various pedoclimatic zones of Moldova (Table 5.6). It has been established that the application of N_{60} to gray wheat with very low nitric nitrogen content (12 kg/ha) has tripled the winter wheat crop. Each kilogram of N applied as additional nutrition was recovered with 39 kg of berries. Nitrogen fertilizers applied on the eroded moderately ordinary chernozem with very low mineral content contributed to the increase of the wheat production by 1.9 times.

Each kilogram of N was recovered with 26 kg of berries (Andrieş et al. 2016). When cultivating this crop on the leached chernozem with moderate $N_{mineral}$ content (about 60 kg/ha), the crop yield was 800 kg/ha and each kilogram of N was recovered with 12 kg of grain. On the same chernozem subtype, with a high N_{min} content (over 140 kg/ha), a high yield of autumn wheat of 5.4 t/ha was obtained, but the plants did not react to the supplementary nutrition with nitrogen.

We will mention that nitrogen fertilizers have helped to increase the gluten content of grains, especially on low-nitrogen-containing soils.

The nitrogen dose (DN) for the expected yield is determined according to the following formula: DN = $[R \cdot C - (N_{rs} + N_{rv} + N_p) \cdot 100]$: K_u, where the symbols used have the following meanings: R— harvest, t/ha; C— consumption (export) of N for the formation of 1 t of autumn wheat, kg; N_{rs}— N_{mineral} reserves in spring soil, kg/ha; N_p— the amount of N accumulated in plants, kg/ha; K_u— coefficient of utilization of nitrogen from fertilizers, %. The methodology of optimizing the mineral nutrition of autumn wheat and the normative basis are set out in previous publications (Andrieş 2007, 2011).

Table 5.6 Reaction of autumn wheat to the application of nitrogen fertilizers depending on the mineral nitrogen reserves in the soil

Mineral Nitrogen Reserve in the Soil	*Number of Experiences*	*Variant*	*Harvest, t/ha*	*Increase in Harvest, t/ha*	*The Content of Gluten in Grains, %*
Gray soil in Cuhureşti village, Floreşti district: humus – 2.2%; mobile phosphorus and exchangeable potassium – optimum, precursor – sunflower					
Very low, 12.0	1	PK	1.2	–	15.1
		PK+N_{60}	3.5	2.3	13.4
		PK+N_{120}	4.9	3.7	23.0
DL (0.95) t/ha = 0.4; P = 2.9%					
Ordinary chernozem moderate eroded in the village of Negrea, Hânceşti district: humus – 2.2%; mobile phosphorus and exchangeable potassium – optimum, precursor – corn for grains, wheat					
Very low, under 35.0	2	PK	1.9	–	16.6
		PK+N_{60}	3.6	1.7	16.8
		PK+N_{120}	3.5	1.6	21.5
DL (0.95) t/ha = 0.3–0.5; P = 2.0–4.0%					
Leached chernozem from the village of Ivancea, Orhei district: humus – 3.6%; mobile phosphorus and exchangeable potassium – optimum, precursor – peas					
Moderate, 60.0	6	PK	3.7	–	26.2
		PK+N_{60}	4.4	0.8	26.4
Leached chernozem from the village of Ivancea, Orhei district: humus – 3.6%; mobile phosphorus and exchangeable potassium – optimum, precursor – peas					
High, over 140.0	3	PK	5.4	–	29.7
		PK+N_{60}	5.5	0.1	30.5
DL (0.95) t/ha = 0.3–0.4; P = 2.8–3.2%					

Source: Data from Andrieş Serafim et al. "Măsuri şi procedee agrochimice de sporire a fertilităţii solurilor erodate". 2016: 51–54.

From the analysis and generalization of the results of the scientific researches carried out in the last 40–50 years in the long-term field experiments, with the application of the fertilizers in the ashes (Andrieş 2007, Андрианов et al. 1989, Андриеш et al. 1993, Боинчан 1999, Громов 2007), the following conclusions can be drawn:

- The soil fertilization mineral system, which has been practiced in the last 20–25 years, can provide an increase in the nutrient content of the soil and crop yield of 25–35%.

However, under the conditions of this soil maintenance system, the organic matter circuit is open, the balance of humus and organic nitrogen is negative. As a result, the soil production capacity in a multiannual cycle will decrease, and the risk of nitrate pollution will be preserved. Contradictions between agricultural chemistry and the maintenance of ecological balance can be reduced by applying fertilizers as recommended in use and by controlling soil fertility. Organomineral soil fertilization system with application of 10–15 t/ha of manure, vegetal debris, calculated mineral fertilizer doses in the area crops (with a share of 20–25% leguminous crops including 10–15% perennial herbs) ensures a balanced balance of organic matter and nitrogen, phosphorus, potassium in the soil and the expected crops. Biologization development of agriculture and livestock (with the respective quantities of organic fertilizer production), a closed-circuit leads to the formation of organic matter and nitrogen in the soil and protection of the environmental degradation and pollution.

Heavy hydrolyzable organic nitrogen, which is extracted from the soil with a 5N H_2SO_4 solution, is a less stable fraction than that of non-hydrolyzable nitrogen. According to the existing concepts (Шконде 1965, Щербаков 1968, Кордуняну et al. 1997), this fraction is the most distant backup to ensure the plant accessible nitrogen in the form. This fraction characterizes the potential reservoir to provide the soil with nutrient nitrogen. Hardly hydrolyzable organic nitrogen complements the reserves of readily hydrolyzable organic nitrogen; therefore, it is the key fraction in the process of transformation of total nitrogen into the soil.

Hardly hydrolyzable organic nitrogen not only complements the reserves of mobile forms but in its organic structure synthesis of more stable organic compounds occurs. Studies carried out between 1973–1979 have shown that poorly hydrolyzable organic nitrogen is represented by various amines, proteins, amino acids and other quite stable nitrogen compounds (Donos 1994a, 1994b, 1995, 1997, 1998a,b, 2001). Their total amount, expressed by the heavy hydrolyzable organic nitrogen, is the highest in the leachate chernozem (Table 5.7). Here, in the 0–40 cm layer, there is a reserve of 1,500–1,600 kg/ha of hardly hydrolyzable organic nitrogen. In the ordinary chernozem, this reserve is 1,100–1,150 kg/ha, and in the gray soil is 1,100–1,200 kg/ha.

The relatively low hydrolyzable nitrogen content is 14–15% of the total in the gray soil, 16–18 in the leachate chernozem and 14–15% in the total nitrogen of the ordinary chernozem (Table 5.7). Research has shown that the systematic application of N_{120} mineral fertilizers to a $P_{60}K_{60}$ backbone contributes to increasing the amount of hydrolyzable heavy nitrogen in the 0–40 cm layer of gray soil and leached chernozem by 5–7% compared to the control, and in ordinary chernozem by 3–5%. This proves that some of the mineral nitrogen in the fertilizer is assimilated by the soil biota and vegetal debris, which as a result of the metabolic processes, is transformed into hardly hydrolyzable organic compounds.

Table 5.7 Average content of organic nitrogen which is difficult to hydrolyzate in different soils of the Republic of Moldova

Thickness of the Soil Layer, **cm**	*Measurement Units*	*The Gray Soil of the Forest*			*Chernozem*				
					Leachate		*Ordinary*		
		1	**2**	**3***	**1**	**2**	**1**	**2**	**3****
0–20	mg/kg	235	253	270	296	321	233	250	254
	% from the total	14.3	14.6	12.0	16.7	16.2	14.3	14.0	11.3
20–40	mg/kg	209	216	190	315	324	224	229	221
	% from the total	15.7	15.4	15.6	18.0	17.6	15.6	15.2	14.7
0–40	mg/kg	1,143	1,208	1,188	1,535	1,619	1,100	1,152	1,296
	% from the total	15.0	15.0	14.0	17.4	16.9	14.9	14.6	15.7

Note: 1–witness–arable land without fertilizers; 2–arable soil with $N_{120}P_{60}K_{60}$; 3*–gray forest soil; 3**–grounded, grounded in 1937.

Source: Data from Donos Alexei, "Acumularea şi transformarea azotului în sol". 2008: 61–62.

Table 5.8 The amino acid content of the organic fraction of the hardly hydrolyzable nitrogen in the soils of Moldova, mg N per 1 kg of dry soil in the layer 0–20 cm

The Aminoacids	*The Gray Soil*		*Chernozem*			
			Leachate		*Ordinary*	
	M	**F**	**M**	**F**	**M**	**F**
Cysteine	8.8	6.6	9.3	6.5	9.3	10.3
Serine	4.4	4.7	4.5	4.4	5.3	4.4
Glutamic acid	5.9	6.3	4.3	4.7	4.8	5.7
Threonine	6.4	5.3	4.2	3.5	3.9	4.6
Alanine	7.7	7.4	16.7	17.2	11.7	10.3
Lysine + isoleucine	5.7	6.4	4.4	5.6	10.3	10.0
Total group a.a.* neutral	**38.9**	**36.7**	**43.3**	**41.9**	**45.3**	**45.3**
Arginine	5.3	5.5	4.0	5.0	3.3	5.3
Lysine	4.5	6.0	5.7	5.2	3.6	3.8
Ornithine	3.8	5.1	4.8	6.8	4.5	3.8
Valine	5.9	6.2	6.5	5.1	6.2	4.8
Total group a.a.* alkali	**19.5**	**22.5**	**21.0**	**22.1**	**17.6**	**17.7**
Asparagine and glutominic acids	**6.9**	**7.5**	**12.2**	**13.0**	**8.7**	**10.0**
Histidine	3.5	4.0	3.7	3.2	2.6	2.8
Tryptophan + phenylalanine	4.5	4.3	6.1	4.7	6.4	6.3
Total a.a.* heterocyclic	**8.0**	**8.3**	**9.8**	**7.9**	**9.0**	**9.1**
Total amino acids	**73.3**	**75.3**	**86.3**	**84.9**	**80.6**	**82.1**

Note: M–unfertilized witness; F–fertilized with $N_{120}P_{60}K_{60}$; a.a.–amino acids.

By determining the content of poorly hydrolyzable organic nitrogen in dynamics, it was found that during the spring period the soil is less than autumn (Donos 2001). Hardly hydrolyzable organic nitrogen is represented by its amine form of protein, and therefore it consists of amino acids, which serve as a "bridge" for the transition of protein substances from one form to another (Table 5.8). Studies (Donos 1994a, 1994b, Donos and Andrieş 2001) have shown that 16 amino acids have been identified and determined in the heavy hydrolyzable nitrogen fraction. The average amino acid content in the hydrolyzable heavy fraction is shown in Tables 5.8 and 5.9.

Table 5.9 Average nitrogen content of amino acids in the organic hydrocarbon fraction of heavy nitrogen

Soil Layer, cm	*Indices of Measurement*	*The Gray Soil of the Forest*		*Chernozem*			
				Leachate		*Ordinary*	
		M	F	M	F	M	F
0–20	mg/kg	70.8	71.5	81.0	83.0	76.6	78.2
	%	30.2	27.7	27.9	25.7	34.4	31.2
20–40	mg/kg	61.5	70.0	73.3	76.6	71.4	74.0
	%	27.9	32.7	23.8	23.6	33.0	34.7
0–40	mg/kg	33.7	36.1	38.7	40.1	35.8	36.6
	%	29.5	29.9	25.2	24.7	32.5	31.8

Source: Data from Donos Alexei, "Acumularea şi transformarea azotului în sol". 2008: 65–66.

The Easily Hydrolyzable Organic Nitrogen in the Soils of Moldova

A component of the total nitrogen in the soil is the nitrogenous compounds of the easily hydrolyzable organic fraction which are extracted with dilute sulfuric acid (0.5 N). Extracts such as amides, aminosugars, amino acids and other peptide derivatives (Donos 2001) are separated into the extract. Studies have shown that slightly hydrolyzable organic nitrogen is less strongly bound to humid substances in the soil and is more prone to mineralization processes than other organic nitrogen fractions (Donos 2008). Easily hydrolyzable organic nitrogen, being comparatively fast undergoing mineralization processes, serves as a reserve for nitrogen plant nutrition (Donos 2006, Donos and Andrieş 2001).

The easily hydrolyzable organic nitrogen constitutes 11–13% of total nitrogen or absolute values of 840–920 kg/ha of nitrogen in the layer of 0.4 m of gray forest soil. In the ordinary chernozem of southern Moldova, the content of the easily hydrolyzable organic nitrogen is 820–870 kg/ha in the 0.4 m soil layer (Donos 2008).

After the absolute amount, the most easily hydrolyzable organic nitrogen among the studied soils was observed in the levigated chernozem. This index, in the 0.4 m layer, averaged 1,108 and 1,175 kg/ha of nitrogen. The least easily hydrolyzable organic nitrogen is contained in the 20–40 cm layer of gray soil (148–169 mg/kg of dry soil), while in the upper layer (0–20 cm) this index is 198–207 mg/kg of soil (Table 5.10). In the soil samples collected from the forest, 247 mg/kg of easily hydrolyzable nitrogen in the 0–20 cm layer was established, and in the 20–40 cm layer only 125 mg/kg. Therefore, the intense and long-lasting use of this type of soil in agriculture leads to the redistribution of easily hydrolyzable organic nitrogen from the upper layer of the soil to the lower one (Table 5.10). However, relatively easily hydrolyzable organic nitrogen content accounts for 10–11% of total nitrogen. In the gray soil used in agriculture, the easily hydrolyzable organic nitrogen content is 12–13% of the total nitrogen (Table 5.10).

Table 5.10 Medium content of easily hydrolyzable organic nitrogen in different soils and categories of use

The Layer of Soil, cm	*Indices of Measurement*	*The Gray Soil of the Forest*			*Chernozem*				
					Leachate		*Leachate*		
		M	F	P	M	F	M	F	Ţ
0–20	mg/kg	198	207	247	229	244	194	203	200
	%	12.7	12.6	11.0	13.0	13.0	12.3	12.2	9.2
20–40	mg/kg	148	169	125	213	225	162	181	155
	%	12.0	12.7	10.3	12.7	12.7	11.9	12.7	10.3
0–40	mg/kg	869	955	926	1,108	1,175	854	922	798
	%	12.3	12.7	10.9	12.8	12.9	12.1	12.4	9.7

Note: %—percentage of total nitrogen; M—unfertilized variant; P—forest; F—fertilized soil $N_{120}P_{60}K_{60}$; Ţ—has not been worked since 1937.
Source: Data from Donos Alexei, "Acumularea şi transformarea azotului în sol". 2008: 67–68.

In the ordinary chernozem left for 38 years, the easily hydrolyzable organic nitrogen content in the 20–40 cm layer was 155 mg/kg of dry soil, which is 10.3% of the total nitrogen. In the upper layer, the absolute content is almost the same as in the soil used in agriculture, however, relative to the total; it is only 9.2% (Table 5.10).

Thus, intensive and long-term soil work contributes to more rapid mineralization of easily hydrolyzable organic nitrogen compounds. This organic nitrogen fraction is more mobile than the hardly hydrolyzable organic nitrogen. Nitrogen compounds of the easily hydrolyzable fraction are characterized as the closest reserves for mineral nitrogen plants (Donos 2008, Кордуняну et al. 1997, Смирнов 1977, Шконде et al. 1964, Шконде 1965).

We note that the systematic application of mineral nitrogen fertilizers ($N_{120}P_{60}K_{60}$) has contributed to the increase of the amount of nitrogen from the easily hydrolyzable fraction by 9–10% compared to the gray soil control variant and by 6–10% in soils of the chernozem type (Table 5.10). Therefore, mineral nitrogen fertilizers activate the biological processes of synthesis of the mobile organic compounds by increasing the vegetal debris, the quantity of which in the fertilized variants was higher.

The amount of easily hydrolyzable organic nitrogen in different crop development periods was different. At the beginning of spring, its content in the soil under autumn wheat is the highest, and in the biological maturity period is the lowest. In plantless variants (black field), the easily hydrolyzable organic nitrogen content at all sampling steps was 5–9% higher than plant variants. This confirms that part of the easily hydrolyzable organic nitrogen is actively involved in plant nutrition processes. Analog data were also obtained in the cultivation of maize (Donos 2008).

Of all the organic compounds from which the easily hydrolyzable nitrogen fraction is common, the amino acids represent on average 21–28% of the total nitrogen of this fraction (Donos 2008). From the soil extract to the easily hydrolyzable organic nitrogen based on the chromatographic paper method (Турчин 1960), we detected 16 amino acids. Studying the quantitative dynamics of amino acid nitrogen, it was established that their peak in soil coincided with the autumn wheat flour and the 6–8 leaf corn phase (Donos 2008).

The second group is the alkali amino acid group (4 in number) with lower total nitrogen content from 10 to 14 mg/kg of dry soil (Table 5.11).

Systematic data in Table 5.9 that the neutral amino acid group consists of 7, the total nitrogen content thereof is 18–22 mg/kg in the gray soil, 28–30 mg/kg in the leachate chernozem and 21–28 mg/kg in the usual one.

Within the amino acids found, the least amount of nitrogen is contained in the heterocyclic amino acid (tryptophan and phenylalanin), constituting only 2.0–4.4 mg/kg of soil (Table 5.11).

Table 5.11 Nitrogen content in amino acids of the easily hydrolyzable organic fraction of different soils (averaged over 1974–1978), mg/kg of dry soil

Name of Amino Acids	*Soil Layer, cm*	*The Gray Soil*		*Chernozem*			
				Leachate		*Leachate*	
		M	F	M	F	M	F
Cysteine	0–20	3.9	3.6	5.6	5.4	4.1	4.7
	20–40	3.2	3.9	6.1	5.0	3.8	3.3
Serine	0–20	3.2	3.4	5.1	4.4	5.0	5.4
	20–40	2.9	3.3	4.1	4.4	4.9	4.8
Glutamic acid	0–20	3.9	3.1	5.2	4.5	3.4	3.0
	20–40	2.8	3.0	5.0	5.5	2.5	2.4
Threonine	0–20	3.7	3.2	4.9	5.4	6.9	5.5
	20–40	3.3	3.4	5.0	5.3	4.4	4.8
Alanine	0–20	4.1	4.3	5.1	4.6	5.0	3.9
	20–40	3.8	4.8	4.6	5.1	4.1	3.8
Leucine + isoleucine	0–20	3.2	2.3	4.6	4.3	3.5	2.5
	20–40	2.4	2.9	4.0	4.3	2.4	2.7
Arginine	0–20	4.1	3.9	4.2	3.9	3.2	2.7
	20–40	3.2	4.0	3.9	4.0	2.9	2.2
Lysine + histidine	0–20	2.5	4.0	4.7	4.0	2.3	2.4
	20–40	3.2	2.6	4.0	3.3	1.9	2.0
Ornithine	0–20	2.3	2.2	5.1	4.8	2.7	2.8
	20–40	2.0	2.9	4.8	5.4	3.0	2.7
Valine	0–20	3.0	3.0	–	–	2.8	3.0
	20–40	3.0	2.5	–	–	2.4	2.8
Asparagine and glutaminic acids	0–20	5.5	5.6	9.9	9.6	9.9	7.8
	20–40	4.8	5.6	9.4	9.7	8.6	6.5
Cyclic amino acids: tryptophan and phenylalanine	0–20	2.1	2.0	4.3	3.8	4.4	3.7
	20–40	2.3	2.2	3.6	4.3	3.3	3.7
In total	**0–20**	**41.5**	**40.6**	**58.7**	**54.7**	**53.2**	**47.4**
	20–40	**36.9**	**41.1**	**54.5**	**56.3**	**44.2**	**41.7**

Note: M–control; F–fertilized with $N_{120}P_{60}K_{60}$.
Source: Data from Donos Alexei, "Acumularea şi transformarea azotului în sol". 2008: 69–70.

After the total nitrogen content of the amino acids of the easily hydrolyzable organic fraction, the leachate chernozem is 55–59 mg/kg of soil. In the ordinary chernozem, this index is 42–53 mg/kg, and in the gray soil – 37–41 mg/kg of dry soil (Table 5.11).

The total nitrogen content of these four groups of amino acids of the organic fraction of the easily hydrolyzable nitrogen fraction is shown in Table 5.12. The neutral amino acid group of the easily hydrolyzable nitrogen constituted from 21 to 25 mg/kg or from 11 to 15% of total nitrogen. In the levigated chernozem, the absolute amount of total nitrogen of this amino acid group constituted 28–31 mg/kg or 12–13.5% of the total nitrogen of the easily hydrolyzable organic fraction.

The ordinary chernozem, after the relative amount of nitrogen of the neutral amino acids, is higher than in the other soils (Table 5.12).

The group of alkaline amino acids in the easily hydrolyzable organic fraction of nitrogen makes up 4–6% of the total nitrogen of the fraction of the acidic 3–5% and the heterocyclic amino acids are only 1–2%.

Table 5.12 Nitrogen content of amino acid groups in the organic fraction of easily hydrolyzable nitrogen in different soils (generalized data)

Amino acid Groups (a.a.)	*Soil Layer,* cm	*The Gray Soil*		*Chernozem*			
				Leachate		*Ordinary*	
		1	2	1	2	1	2
Neutral	0–20	25.0 / 12.6	22.9 / 11.1	30.9 / 13.5	28.6 / 11.7	30.7 / 15.8	28.0 / 13.8
	20–40	21.9 / 14.8	23.8 / 14.1	28.8 / 13.5	29.6 / 13.2	24.5 / 15.1	24.6 / 13.6
Basic	0–20	8.9 / 4.5	10.1 / 4.9	14.0 / 6.1	12.7 / 5.2	8.2 / 4.2	7.9 / 3.9
	20–40	8.4 / 5.7	9.5 / 5.6	12.7 / 6.0	12.7 / 5.6	7.8 / 4.8	6.9 / 3.8
Acidic	0–20	5.5 / 2.8	5.2 / 2.5	9.9 / 4.3	9.6 / 3.9	9.9 / 5.1	7.8 / 3.8
	20–40	4.8 / 3.2	5.6 / 3.3	9.4 / 4.4	9.7 / 4.3	8.6 / 5.3	6.5 / 3.6
Cyclic	0–20	2.1 / 1.1	2.0 / 1.0	4.3 / 1.9	3.8 / 1.6	4.4 / 2.3	3.7 / 1.8
	20–40	2.3 / 1.6	2.2 / 1.3	3.6 / 1.7	4.3 / 1.9	3.3 / 2.0	3.7 / 2.0
In total	0–20	41.5 / 20.9	40.6 / 19.6	58.7 / 25.8	54.7 / 22.4	53.2 / 27.4	47.4 / 23.3
	20–40	36.9 / 24.9	41.1 / 24.3	54.5 / 25.6	56.3 / 25.0	44.2 / 27.3	41.7 / 23.0

Note: to the denominator–in mg/kg of dry soil; in the numerator in % of the easily hydrolyzable organic nitrogen; 1–unfertilized variant; 2–fertilized with $N_{120}P_{60}K_{60}$.
Source: Data from Donos Alexei, "Acumularea şi transformarea azotului în sol". 2008: 71–72.

Thus, the amino acids in the easily hydrolyzable organic nitrogen fraction represent over 20% of the total nitrogen of the fraction, which is actively involved in the processes of transformation (synthesis and mineralization) of the total nitrogen in the soil. This conclusion is also substantiated by the fact that in the period of growth and development of crops, the easily hydrolyzable organic nitrogen content changes the quantitative and qualitative part, first of all, of the share of amino acids (Donos 2008), which is part of the metabolism process of organomineral substances in the soil.

In all the studied soils it was observed that the reserve of easily hydrolyzable organic nitrogen during the spring period was the highest, and during the harvesting period of the autumn wheat and maize were the lowest.

It should be noted that among the soils investigated, the most easily hydrolyzable organic nitrogen is contained in the levigated chernozem. Following the easily hydrolyzable nitrogen content, the differences between gray soil and ordinary chernozem are insignificant. Variants fertilized with N_{120} $(PK)_{60}$ contain more easily hydrolyzable nitrogen than unfertilized nitrogen. The dynamism of the easily hydrolyzable organic nitrogen was studied not only in different soils and stages of the development of cereal crops without and with the application of fertilizers but also in soil without plants (black field). This allowed us to determine the proportion of easily hydrolyzable organic nitrogen in providing accessible nitrogen crops.

Referring briefly to these investigations, we can see that:

a) In the gray soil, the easily hydrolyzable organic nitrogen reserve during the spring period represents in the 0.4 m layer about 1,000 kg/ha, and when the wheat is harvested it is only 805 kg/ha; in maize, it is 879 kg/ha in spring and 755 kg/ha in autumn (Table 5.13). In the variants without plant cultivation, the difference between spring and autumn is lower than in plant variants (54 and 60 kg/ha). So, maize plants have used 60 kg/ha lightly-nitrogen-free nitrogen feed, and 54 kg/ha of autumn wheat (Table 5.13). This constitutes 5–7% of the easily hydrolyzable organic nitrogen reserve, which is determined during the spring period.
b) In the levitated chernozem, the difference in the easily hydrolyzable nitrogen content between the spring and the wheat harvest period is 68 kg/ha; in the cultivation of maize, it is 106 kg/ha (Table 5.13), which represents 5.9% and 9.4% of the easily hydrolyzable nitrogen stock in the 0.4 m layer during the spring period.
c) In the ordinary chernozem, the difference between the spring and autumn content of the easily hydrolyzable organic nitrogen stock is 135 kg/ha in the cultivation of autumn wheat is 79 kg/ha (Table 5.13). These values are equivalent to 13.3% of the easily hydrolyzable nitrogen reserve for wheat cultivation, and 8.7% to the cultivation of maize from its reserve in the spring.

The easily hydrolyzable organic nitrogen content may serve as an acceptable indicator for the assessment of crop assurance with nutrient nitrogen and as an index for the prediction of the use of nitrogen fertilizers.

We note that these quantitative changes in the organic fraction of the easily hydrolyzable nitrogen occur especially due to the monoaminocarbonate, diaminocarbonate and monoaminodicarbonate of amino acids (Donos 2008).

The share of their participation in processes of mineralization of easily hydrolyzable nitrogen during the vegetation period constitutes from 5 to 15% of the total readily hydrolyzable nitrogen of this organic fraction. Therefore, the easily hydrolyzable organic nitrogen fraction is the closest source for providing nutrient nitrogen crops. Its total content during the spring period serves as an index of agricultural crop insurance with accessible nitrogen.

Research on total nitrogen components has shown that the percentage of organic compounds is 94–97% of the total. These organic compounds were divided into non-hydrolyzed compounds with concentrated sulfuric acid solutions, hardly hydrolyzed in 5N H_2SO_4 solutions and readily hydrolyzed in 0.5N H_2SO_4 solutions.

Table 5.13 Change of easily hydrolyzable organic nitrogen contents in the soil layer of 40 cm in the cultivation of cereal crops on various soils, kg/ha

Variant	*The Gray Soil of the Forest*			*Chernozem*					
				Leachate			*Leachate*		
			The Difference			*The Difference*			*The Difference*
	1	2	1–2	1	2	1–2	1	2	1–2
				Autumn Wheat					
Without plants (black field)	995	855	140	1,172	1,103	69	1,009	988	21
With plants	999	805	194	1,145	1,008	137	1,015	859	156
The difference ±	−4	+50	−54	+27	+95	−68	−6	+129	−135
In % of 1	–	–	5.4	–	–	5.9	–	–	13.3
				Maize for Grains					
Without plants (black field)	897	833	64	1,135	1,055	80	925	847	78
With plants	879	755	124	1,130	944	186	912	755	157
The difference ±	+18	+78	−60	+5	+111	−106	+13	+92	−79
In % of 1	–	–	6.8	–	–	11.2	–	–	8.7

Note: 1–the spring period; 2–cropping period.
Source: Data from Donos Alexei, "Acumularea şi transformarea azotului în sol". 2008: 73–74.

Non-hydrolyzable nitrogen is 65–70% of the total in arable soils, and 76% of the total is used in the gray soils of forest and ordinary chernozem. The systematic application of mineral fertilizers with nitrogen has increased the non-hydrolyzable nitrogen reserve by 3–5% compared to non-fertilizer variants.

Hardly hydrolyzable organic nitrogen is a remote reserve of access to affordable nitrogen plants and represents 14–18% of the total in arable soils and 11–12% of the total in the gray soils of forest and ordinary chernozem. The systematic application of mineral fertilizers with nitrogen has increased the reserve of hydrolyzable nitrogen by 4–5% compared to unfertilized variants. The composition of the hard hydrolyzable organic nitrogen was identified and determined 16 amino acids, which together account for 28–32% gray ground, the levigated was typically 23–28% and chernozem was total nitrogen 31–35% of the fraction.

The easily hydrolyzable organic nitrogen is the closest reserve for providing accessible nitrogen plants and constitutes 11–13% of the total nitrogen of soils in arable soils. In the forest and unworked soils, the easily hydrolyzable nitrogen content is 2–3% lower than in arable soils. The systematic application of nitrogen fertilizers increased the easily hydrolyzable nitrogen reserve by 6–9% compared to unfertilized variants. Of the nitrogen-containing organic compounds composed of the easily hydrolyzable nitrogen fraction, the amino acids constitute 21–28% of the total nitrogen of the fraction.

The nutrients the plant takes from the soil, nitrogen is the first: the percentage of plant nitrogen and its removal from crops is greater than any other element. The higher the yields, the large natural nitrogen quantities are estranged from the soil.

In order to obtain high yields across crops, even on soils rich in humus and nitrogen, the amount of mineral nitrogen accumulated by mineralization processes; in other words, the mobilization of their natural reserves even if are improved by techniques processing is not sufficient enough.

Among the cultural effects of nitrogen enrichment, vegetable culture, thanks to symbiosis with nodule bacteria, uses nitrogen in the atmosphere, is of paramount importance. The higher the leguminous production, the higher the nitrogen content of the soil. The distinct types of vegetables, whose culture pursues different economic objectives, have a different effect on soil nitrogen reserves. Associated nitrogen accumulated in leguminous plants greatly increases the soil stock if the beans are shed in green fertilizers (lupins).

The nutrients that the plant takes from the soil, nitrogen takes the first place: the percentage of nitrogen in plants and connection with this removal of it from the crops is higher than any of the other elements. The higher the yields, the naturally large amounts of nitrogen are alienated from the soil.

To obtain high yields from the entire area of crops, even on soils rich in humus and nitrogen, the amount of mineral nitrogen that accumulates in them as a result of mineralization processes; in other words, the mobilization of their natural reserves, even if they are enhanced by processing techniques, is not enough.

Of the cultural effects aimed at enriching the soil with nitrogen, the culture of legumes, thanks to the symbiosis with nodule bacteria, uses nitrogen from the atmosphere, is of paramount importance. The greater yield of legumes, the higher their value in enriching the soil with nitrogen. Separate types of legumes, whose culture pursues different economic goals have a different effect on the nitrogen reserves in the soils. Associated nitrogen accumulated in leguminous plants most fully increases the soil stock in the case of plowing bean to green fertilizer (lupine).

When cultured leguminous grasses in the hay in the soil, only nitrogen of root and post-plant residues are preserved, which is 1/2 of the total nitrogen of plants in clover (75–80 kg/ha N per year) and 1/3 of nitrogen in alfalfa (100 kg/ha N per year) The rest of the nitrogen along with the hay is removed from the field and can be returned to the soil only in the form of manure from the livestock that is fed hay. Grain legumes (peas, beans, soybeans and beans), giving a protein-rich grain (which has great feed and nutritional value), contain in the aboveground mass (mainly in grain) a large amount of nitrogen accumulated due to the activity of nodule bacteria, which not only do not enrich but even somewhat impoverish the soil with nitrogen (Donos 2008). Nitrogen grain legumes can return to the soil only with manure.

The use of manure and other organic fertilizers (peat and composts) is the most universal way to restore nitrogen reserves (as well as other nutrients in the soil). Manure on average contains 0.5% N; therefore, about 100 kg of nitrogen is applied from 20 tons of manure per hectare. But the actual amount of nitrogen that enters the soil with manure largely depends on how it is prepared, stored, applied and embedded in the soil.

Together with the development of the chemical industry, the use of mineral nitrogen fertilizers is becoming increasingly important. The most common currently are ammonium nitrate, ammonium sulfate and urea. These are highly soluble salts, the ammonium of which is vigorously absorbed by the soil; liquid ammonia is also of great importance.

To fulfill the tasks currently facing agriculture in our country, the intensive use of soil, obtaining maximum yield per hectare, expansion of areas under highly productive crops that place high demands on the nutritional regime, while maintaining and increasing soil fertility requires extensive use of all ways to replenish nitrogen reserves in the soil.

The introduction of a large amount of nitrogen in the form of nitrogen fertilizers has a very strong influence on the dynamics of nitrogen in agricultural systems. The introduction of nitrogen into the soil increases productivity and contributes to the accumulation of biomass in the short term. Thus, increasing the level of nitrogen in the soil was perceived as a strategy aimed at preserving soil carbon. However, the use of nitrogen as a fertilizer is accompanied by the cost of CO_2 emissions caused by the production, packaging, transportation and fertilization. In addition to this, increasing the content of soil organic matter can accelerate the dynamics of nitrogen accumulation and of course the release of N_2O as a greenhouse gas.

Thus, nitrogen affects the balance of greenhouse gases in four ways:

1. CO_2 is released from the fuel required for intensive production of nitrogen fertilizers.
2. Changes in crop yields depending on the rate of nitrogen input. An increase in the use of nitrogen fertilizers may lead to a decrease in soil pH. Neutralization of the acidic soil environment requires the use of lime, the production of which is also an energy-intensive process and is accompanied by the release of CO_2.
3. N_2O emissions vary depending on the tillage technology and the rate of applied nitrogen.
4. Technologies that allow controlling the carbon content in the soil: conventional technologies of tillage or technologies of soil-protective and resource-saving agriculture.

Natural Environment and Efficacy of Fertilizers

The territory of the Republic of Moldova is characterized by a rough relief. The predominance of slopes on 80% of the territory creates favorable conditions for the development of erosion processes. The average absolute altitude of the Republic of Moldova is 147 m, the maximum is 429 m and the minimum is 5 m. The surface of the eroded soils, which lost from 20 to 70% of their initial fertility, represents about 36% (Cadastru Funciar al Republicii Moldova 2012). According to the Statistical Yearbook of Moldova (Anuarul Statistic al Moldovei, 2012), on January 1, 2012, the total land area was 3.38 million hectares, including agricultural land of 2.5 million ha (73.8%) and the forestry fund was 463.1 thousand ha (13.7%). Of the total area of 2.5 million hectares of agricultural land, arable lands constitute 1.81 million ha (72.6%), orchards is 133.3 thousand (5.3%), vineyards is 149.6 thousand ha (6.0%) and pastures is 350.4 thousand ha (14.0%).

According to the data presented, the share of agricultural land is inadmissible (73.8%) and the forestry fund is 2–3 times lower than the optimal one. The imbalance between natural and anthropogenic ecosystems conditional the amplification of various forms of soil degradation is observed.

The climate of Moldova is temperate continental (Лаce 1978) with mild and short winter (average temperature in January is 3–5°C), hot and long summer (average temperature in July plus 20–22°C). In relation to the climatic index values, the territory of Moldova was divided into three zones, which are also pedoclimatic: North, Central and South. The amount of atmospheric precipitation varies between 500–630 mm in the North and 450–500 mm in the South (Агроклиматические ресурсы Молдавской ССР 1982). The sum of temperatures above 100°C is 2,750–28,500°C in the Northern Zone and 3,100–33,500°C in the South. The hydrothermal coefficient (K after Ivanov–Vişotchi) in the North of the country is 0.7–0.8 and in the South is 0.5–0.6. The frequency of droughts in ten years is once in the North, 2–3 times in the Center and 3–4 times in the South.

The structure of the soil cover is quite complex. The main types and subtypes of soil are chernozems occupying 70%, brown and gray soils occupying 10.2%, alluvial soils are 10.2% and delluvial soils are 4.0% (Крупеников and Подымов 1987, Почвы Молдавии 1984, Program complex de valorificare a terenurilor degradate şi sporirea fertilității solurilor. Partea II. 2004b). Soils with high fertility and favorable heat regime allow cultivating a wide spectrum of valuable crops, such as vines, oleaginous crops, fruit trees, nuts, vegetables, sunflowers, etc., producing high quality. The current state of soil cover quality is presented in Table 5.14. Soils with a score of 80–100 points account for about 27% of the total area of agricultural land (Program complex de valorificare a terenurilor degradate şi sporirea fertilității solurilor. Partea II. 2004b).

On these highly productive soils, usually presented by typical and leached chernozems (standard soils) with an organic matter content of 3.6–4.5%, it can be obtained from the actual fertility rate of 3.2–4.0 t/ha of autumn wheat. Areas of Class II and III ranging from 60–80 points occupy 36% or 918 thousand ha. The productivity of these soils is also quite high and represents 2.4–3.2 t/ha of autumn wheat.

The soils of these two grades are frequently affected by processes of humus loss, lack of nutrients, destruction and secondary compaction and biological degradation partly due to surface erosion. Soils in grades IV to VI account for 30% of the total, have a score of 20–60 points and low productivity of 0.8–2.4 t/ha of winter wheat. These soils are weak, moderate and strongly degraded, especially due to erosion.

At present, according to the Land Cadastre of the Republic of Moldova on January 1, 2012 (Cadastru Funciar al Republicii Moldova 2012), the credit score represents 63 points in the republic. Effective soil fertility ensures the formation of 2.5 t/ha of autumn wheat. Under Moldova conditions, soil humidity (atmospheric precipitation) is one of the natural factors that determine the formation of high and stable crops.

The calculations made by the Institute of Pedology, Agrochemistry and Soil Protection "Nicolae Dimo" (Program complex de valorificare a terenurilor degradate şi sporirea fertilităţii solurilor. Partea II. 2004b) have demonstrated that in a multiannual cycle the average winter harvest of precipitations 4.3 t/ha. The harvest difference is based on the amount of rainfall and the credit score is high and makes (4.3–2.5 t) 1.8 t/ha. Given the insufficiency or deficiency of nutrients, the unsatisfactory condition of the physical and biological properties of the soil, the plants unproductively consume the moisture reserves accumulated in the soil for the synthesis of organic substances as a result the crops are low and of low quality.

Table 5.14 The state of quality (bonitet) of the soil cover of the Republic of Moldova

The Creditworthiness Class	*Note Creditworthiness, Points*	*Percentage of the Surface of Agricultural Goods*	*Surface, Thousands of* **ha**	*Harvest of Autumn Wheat,* **t/ha**
I	81–100	27	689	3.2–4.0
II	71–80	21	539	2.8–3.2
III	61–70	15	382	2.4–2.8
IV	51–60	15	382	2.0–2.4
V	41–50	9	303	1.6–2.0
VI	1–40	6	153	0.8–1.6
VII	> 20	7	178	–
Average on the Republic	65	100	2556	2.6

Source: Data from "Program complex de valorificare a terenurilor degradate şi sporirea fertilităţii solurilor". 2004: 42–105.

These are confirmed by research conducted in long-term field experiments. It has been established that in optimally fertilized variants, crop plants consume 20–25% less water compared to unfertilized variant (Andrieş 2007).

I. Krupenikov (Крупеников 2008), analyzing the main forms of degradation of chernozems (total 11 forms), first mentioned the humic degradation and in the second mentioned the agrochemical degradation (soil shedding in nutrients). These two forms of degradation occur permanently on all agricultural lands.

The multiannual results obtained in the field experiments showed that, under the conditions of the Republic of Moldova, the application of optimal fertilizers provides a 66% increase in sugar beet, 48% in winter wheat and 35% in the cultivation of grain maize and flower (Table 5.15).

The productivity of crop plants in fertilized variants was 4.3 t of autumn wheat, 5.4 t maize for grains, 2.0 t of sunflower seeds and 34.8 t/ha of sugar beet (Andrieş 2007).

In the Republic of Moldova there were elaborated norms for the determination of the need of fertilizers for obtaining the expected crops (Нормативы по использованию минеральных и органических удобрений в сельском хозяйстве Молдавской ССР 1987). It has been established that optimal fertilizer application results in a significant increase in the harvest of 1.2 tons of autumn wheat, 1.4 t maize for grains, 13.8 t of sugar beet and 0.5 t/ha of seeds of sunflower.

From the data presented, soil fertilization and the optimization of the mineral nutrition of crop plants are important factors for obtaining high yields.

Use of fertilizers and balance of nutrients in soil. The soils of the Republic of Moldova are characterized by a high fertility (Почвы Молдавии 1984, 1986, Program complex de valorificare a terenurilor degradate şi sporirea fertilităţii solurilor. Partea I şi II. 2004a,b, Крупеников 1967, 1992, Cerbari 2010, Ursu 2011). The researches carried out in the 1950s and 1960s showed that the chernozems of Moldova contained 340 t/ha of humus in the 100 cm layer during this period.

Table 5.15 Efficiency of fertilizers under the conditions of the Republic of Moldova

Culture	*Harvest, t/ha*		*Increase in Harvest*	
	Unfertilized Soil	*Fertilized Soil*	**t/ha**	*%*
Autumn wheat	2.9	4.3	1.4	48
Maize for grains	4.0	5.4	1.4	35
Sunflower	1.5	2.0	0.5	35
Sugar beet	21.0	34.8	13.8	66

Source: Data from Andrieş Serafim, "Optimizarea regimurilor nutritive ale solurilor şi productivitatea plantelor de cultură". 2007: 73–248.

The composition of organic matter contained about 20 t/ha of nitrogen and 5 t/ha of phosphorus. The total amount of P_2O_5 was about 160–180 mg in the layer, and at a depth of 90–100 cm was up to 100 mg in 100 g of soil. The total phosphorus reserve in the 1 m layer was 17 t/ha. The soils of Moldova are rich in minerals containing potassium (Алексеев 1999). Their overall content is 10–14%. The total potassium reserve in the 1 m layer of chernozems is 170–290 t/ha.

Between 1950 and 1960, the crops were quite modest and consisted of 1.6 t of autumn wheat, 2.8 t of maize for grains, 1.5 t of sunflower seeds and 19.2 t/ha of sugar beet (Table 5.16). High yields were limited by two natural factors: 1) insufficient moisture and 2) low levels of nutrients in the soil. Possible harvests calculated by the degree of moisture insurance were 60–70% higher than those obtained at that time (Table 5.17).

This made it possible to assume that first of all limiting factors belonged to the lack of nutrients in the soil.

Generally, the efficacy of fertilizers (E) is expressed by the equation:

$$E = Ra.a. - Rs.n. \quad (1)$$

in which

Ra.a.– the size of the harvest is limited by the degree of water supply;

Rs.n.– the size of the crop is determined by the nutrient content of the soil.

Until 1965, the beginning of fertilizers in Moldovan agriculture was insignificant. According to statistical data, 6.2 kg of N, 8.7 kg of P and 3.6 kg/ha of K with mineral fertilizers were introduced into 1 ha of arable land and vineyard plantations during the years 1961–1965. The average organic fertilizer dose was 1.3 t/ha (Table 5.18).

Table 5.16 The dynamics of the main crop yields in Moldova, t/ha

Years	*Autumn Wheat*	*Maize for Grains*	*Sunflower*	*Sugar Beet*
1963–1965	1.6	2.8	1.5	19.2
1966–1970	2.0	3.4	1.6	25.6
1971–1975	3.4	3.6	1.8	27.9
1976–1980	3.5	3.6	1.7	27.8
1981–1985	3.4	2.7	1.8	28.7
1986–1990	3.8	3.9	2.0	24.8
1991–1995	3.5	2.7	1.4	24.8
1996–2000	2.1	3.0	1.1	19.0
2001–2005	2.2	2.8	1.2	22.7
2006–2010	2.2	2.7	1.3	27.1

Source: Data from Andrieş Serafim, "Optimizarea regimurilor nutritive ale solurilor şi productivitatea plantelor de cultură". 2007: 33–189.

Table 5.17 Forecast of field crop yields according to the degree of water supply, t/ha

Culture	*Water Consumption for* **1** *Ton of Production,* **t**	*Moisture Reserve in Soil (by zone),* **t/ha**		
		North	*Center*	*South*
		4,010	3,620	2,920
		Harvest, t/ha		
Autumn wheat	820	4.9	4.4	3.6
Maize for grains	640	6.3	5.6	4.7
Sunflower	1,330	3.0	2.7	2.2

Source: Data from Andrieş Serafim, "Optimizarea regimurilor nutritive ale solurilor şi productivitatea plantelor de cultură". 2007: 50–77.

Exports of soil nutrients to crops were significant. As a result, a deeply deficient balance of nutrients was formed in Moldovan agriculture. During the examined period the nutrient deficit per hectare was 59 kg of N, 14 kg of P_2O_5 and 80 kg of K_2O (Table 5.19).

Research in the years 1955–1970 showed that fertilizers were effective in all crops and on all soils (Andrieş et al. 2000, Почвы Молдавии 1986).

This fact has conditioned the accelerated pace of agricultural chemistry. The volume of the mineral fertilizers used on the arable land and in the vineyard plantations increased rapidly. In 1970, the agrarian sector of the republic received fertilizers of 4.1 times more compared to 1963. The fertilizer application rate was 62.7 kg/ha of NPK. As a result, the balance of nutrients improved rapidly.

During 1981–1988, for the first time in the history of Moldovan agriculture, the balance of nutrients became positive. During this period, 100 kg N, 66 kg P_2O_5 and 87 kg K_2O were applied to each hectare of arable land and plantations with mineral and organic fertilizers. The average dose of manure used in agriculture was 6.0–6.6 t/ha. As a result, the productivity of crop plants has increased significantly. On average, the autumn wheat harvest in the republic was 3.8 t, maize for grain 3.9 tons and sunflower for 2.0 t/ha.

During the agriculture chemistry, which lasted for 25 years (1965–1990), 1,200 kg of nitrogen, 960 kg of phosphorus and 860 kg of potassium were applied.

Table 5.18 Dynamics of application of industrial and organic fertilizers in the agriculture of Moldova

Years	*Mineral Fertilizers*						Organic Fertilizers, t/ha	*Mineral and Organic Fertilizers*, t/ha *of Arable Land and Vineyard Plantations*		
	Thousands of Tons			kg/ha *Arable Land and Vineyard Plantations*						
	N	P_2O_5	K_2O	N	P_2O_5	K_2O		N	P_2O_5	K_2O
1961–1965	13.0	19.0	8.0	6.2	8.7	3.6	1.3	12.7	12.0	11.4
1966–1970	33.8	3.2	15.4	15.7	15.8	7.2	1.4	22.7	19.3	15.6
1971–1975	75.6	56.0	34.2	35.4	26.2	15.9	2.9	49.9	33.4	33.4
1976–1980	99.6	84.2	59.8	46.6	39.4	27.9	4.1	66.1	50.4	52.5
1981–1985	148.2	102.4	111.4	70.4	48.6	53.0	6.6	101.4	65.1	92.6
1986–1990	76.0	61.0	50.0	36.5	29.3	24.0	3.0	52.0	37.0	42.0
1991–1995	38.0	28.2	13.3	18.8	13.1	6.1	1.8	28.0	17.5	17.2
1996–2000	8.0	0.3	0.1	3.6	0.14	0.04	0.06	4.2	0.4	0.9
2001–2005	13.6	0.6	0.2	4.6	0.1	0.1	0.02	6.5	0.32	0.3
2006–2010	16.1	1.9	1.0	17.7	0.9	0.9	0.02	18.5	2.7	2.0

Source: Data from Andrieş Serafim, "Optimizarea regimurilor nutritive ale solurilor şi productivitatea plantelor de cultură". 2007: 108–160.

Table 5.19 Balance of nitrogen, phosphorus and potassium in the soil of Moldova, kg/ha

Years	N	P_2O_5	K_2O	Sum NPK
1913	–22	–13	–52	–92
1940	–26	–15	–62	–99
1945	–15	–15	–52	–82
1950	–27	–13	–68	–108
1951–1955	–27	–12	–62	–102
1956–1960	–40	–14	–82	–136
1961–1965	–59	–14	–80	–132
1966–1970	–36	–9	–84	–130
1971–1975	–22	–1	–79	–103
1976–1980	–15	+11	–66	–69
1981–1985	+9	+22	–33	–4
1986–1990	–15	+25	–49	–8
1991–1995	–18	–11	–80	–113
1996–2000	–30	–21	–83	–134
2001–2005	–24	–23	–81	–128
2006–2010	–26	–22	–84	–132

Source: Data from Andrieş Serafim, "Optimizarea regimurilor nutritive ale solurilor şi productivitatea plantelor de cultură". 2007: 125–186.

The accumulation of nutrients in the soil was relatively small compared to their export throughout the history of agriculture. Only for 100 years on each arable land were exported 2,300 kg of nitrogen, 1,000 kg of phosphorus and 5,000 kg of potatoes (Zagorcea 1989).

In the years (2006–2018), the volume of mineral fertilizers increased compared to the period 1996–2006, but even the level of 1961–1965 was not reached.

Currently, nitrogen fertilizers are mainly applied, and phosphorus fertilizers—a minimum element in the soils of Moldova—do not apply. In the past 10–20 years, the manure used in Moldova's agriculture constitutes 0.02 t/ha and the optimal norm being about 10 t/ha (Andrieş et al. 2000, Program complex de valorificare a terenurilor degradate şi sporirea fertilităşii solurilor. Partea II. 2004b, Cerbari 2010, Почвы Молдавии, 1986).

In the last years (2005–2018), the average fertilizer standard applied in Moldova's agriculture was 25 kg/ha. From the total dose, about 90–95% is nitrogen fertilizers. The balance of nutrients in the soil is negative (Table 5.19), accelerated chemical degradation of the soil occurs, resulting in low and low yields.

The Needs of the Republic of Moldova in Mineral Fertilizers

Under the conditions of the Republic of Moldova, the natural factors limiting high yields are the lack of nutrients in the soil and the lack of moisture. In order to obtain a 40–50% increase in crops, it is necessary to compensate for nutrient deficiency by application of fertilizers and rational utilization of soil moisture (Andrieş 2007, Andrieş 2011, Andrieş et al. 2012, Cerbari 2010, Program complex de valorificare a terenurilor degradate şi sporirea fertilităţii solurilor. Partea II. 2004b).

In determining the necessity of Moldovan agriculture in fertilizers, the respective decisions of the Moldovan Government, the Ministry of Agriculture and Food Industry on the development of different branches of agriculture by 2020, the statistical data from the last years, the recommendations and norms on the application of fertilizers typical for the pedoclimatic areas of the Republic of Moldova. The optimal level of fertilization provides for the increase of soil fertility, the obtaining of high yields and the maximum profit on a unit of agricultural land, the protection of the environment of nutrient pollution (Program complex de valorificare a terenurilor degradate şi sporirea fertilităţii solurilor. Partea II. 2004b).

The optimal fertilizer application system is provided for a modern level of agriculture with respect to zonal ash, conservative soil conservation, integrated plant protection, the extension of irrigation, zootechnical development and implementation of intensive plant cultivation technologies. This system is based on the combined application of organic and mineral fertilizers to the fullest use of biological nitrogen. Optimal doses of fertilization of the crop plants are shown in Table 5.20.

Fertilizer norms vary by crop from 50 kg for peas to 225 kg/ha NPK for sugar beet. According to the Program (Program complex de valorificare a terenurilor degradate şi sporirea fertilităţii solurilor. Partea II. 2004b), the average annual dose of fertilizers on agropedoclimatic areas is composed of:

- North– 5 t/ha manure and $N_{61}P_{50}K_{20}$;
- Center– 4 t/ha manure and $N_{54}P_{45}K_{18}$;
- South– 4 t/ha manure and $N_{47}P_{43}K_{18}$.

The implementation of crops with an optimal share of leguminous crops will allow the accumulation of 30–35 kg/ha per year from the account of biological nitrogen.

Table 5.20 Optimal doses of mineral fertilizers for the main crops fertilization, kg/ha active substance

Culture	*Recommended Dose*			*Remark*
	N	P_2O_5	K_2O	
Autumn wheat	80	60	40	Annual
Autumn barley	34	60	0	*
Spring barley	34	60	0	*
Maize for grains	60	50	0	*
Peas for grains	30	20	0	*
Sugar beet	105	80	40	*
Sunflower	45	40	40	*
Tobacco	35	40	40	*
Potatoes	60	60	60	*
Vegetables	90	60	60	*
Maize for silo	40	40	0	*
Vineyards bearing fruit	60	60	60	Once every 3 months
Fruit orchards	60	60	60	Once every 3 months
New vineyards (foundation)	–	400	400	Unclog
New orchards (foundation)	–	400	400	Unclog

Source: Data from Andrieş Serafim, "Optimizarea regimurilor nutritive ale solurilor şi productivitatea plantelor de cultură". 2007: 188–294.

The systematic application of mineral and organic fertilizers in doses of $P_{55–60}$ will allow forming in a multiannual cycle a positive balance and an optimal level of phosphorus in the soil to obtain high yields. The average K_{19} fertilizer dose will be insufficient to stabilize potassium in the soil. The compensation for potash losses will be covered from the local fertilizer account and the application of the secondary production as an organic fertilizer. Nitrogen deficiency will be compensated for by biological nitrogen (30–35 kg/ha), manure (24–30 kg/ha) and mineral fertilizer (50–60 kg/ha).

The share of nitrogen in mineral fertilizers will account for about 50% of the total. The optimal need for nitrogen fertilizer crops will be 82.3 thousand t of active substance or N_{55} on average at 1 ha (Table 5.21).

Table 5.21 Annual mineral fertilizer requirements for optimal fertilization of agricultural crops, thousands tons of active substance

Branch, Culture	N	P_2O_5	K_2O
Crops rotation	82.3	69.9	28.4
Vegetables and potatoes	6.8	9.0	6.8
Vineyards bearing fruit	1.5	1.5	1.5
Fruit orchards	2.0	2.0	2.0
New vineyards (foundation)	0	2.1	2.1
New orchards (foundation)	0	1.0	1.0
Additional for irrigated land	6.3	4.6	3.1
Other cultures	1.0	1.0	1.0
Total on the Republic of Moldova	**99.9**	**91.1**	**45.9**

Source: Data from Andrieş Serafim, "Optimizarea regimurilor nutritive ale solurilor şi productivitatea plantelor de cultură". 2007: 237–314.

For cultivation of potatoes and leguminous crops, 6.8 thousand tons N will be needed with the average dose at 1 ha – N_{60}. For the fertilization of fruit orchards, 2.0 thousand tons of nitrogen will be needed for the vineyards 1.5 thousand tons. The need for phosphatic fertilizers will be 69.9 thousand tons for field crops, 9.0 thousand tons for vegetables and potatoes, 1.5 thousand tons for vineyards per fruit and 2.0 thousand tons for fruit orchards. The annual potash requirement will be 28.3 thousand tons for field crops, 6.8 thousand tons for vegetables and potatoes and 3.1 thousand additional tons for irrigated land.

The total annual demand for fertilizers for Moldova after 2020 will be 236.7 thousand tons of active substance, including 99.9 thousand tons of nitrogen, 91.0 thousand tons of phosphorus and 45.8 thousand tons of potassium. This level of fertilization was achieved in the years 1976–1985, applying 243.6–362.0 thousand tons per year (Table 5.18).

The use of the optimal fertilization system coupled with the other technological links of cultivation of crop plants will allow obtaining 4.0–4.2 t of autumn wheat, 5.8 t of maize for grains and a stable balance of nutrients in the agriculture of Moldova.

Primary measures to Conserve and Increase Effective Soil Fertility

For the preservation and enhancement of soil fertility, the collaborators of the Institute of Pedology, Agrochemistry and Soil Protection "Nicolae Dimo" have developed a complex of plant, agro-chemical and agrochemical measures, which includes (Andrieş 2007, 2011, Andrieş et al. 2001, 2012, Program complex de valorificare a terenurilor degradate şi sporirea fertilităţii solurilor. Partea I şi II 2004a,b):

- Optimization of crops and their implementation in each pedoclimatic area.
- Increasing the share of perennial grasses (alfalfa and sparceta) in field crops up to 10–12%.
- Increasing the annual legume crops (peas, beans and soybeans) to about 10–12%.

These changes in the structure of the crops will allow:

- The annual accumulation of about 40–50 thousand tons of biological nitrogen or 30–35 kg/ha.
- Annual soil incorporation of 5–6 t/ha manure; total 9–10 million tons.
- An annual application of 100 thousand tons of nitrogen and 90 thousand tons of phosphorus; total 190 thousand tons.
- Minimization of soil erosion within the permissible limits of about 5 t/ha.

Over the last few years, state programs have been developed for the remediation of soil chemical, physical and biological properties, soil and water pollution protection with nutrients and phytosanitary substances, including:

- Complex program for land degradation and increase soil fertility. Part I.
- Soil Improvement, approved by the Decision of the Government of the Republic of Moldova no. 636 of May 26, 2003.

- Complex program for land degradation and increase soil fertility. Part II. Increasing soil fertility, approved by Government Decision no. 841 of 26 July 2003.
- The Program for the Conservation and Enhancement of Soil Fertility for the years 2011–2020, approved by the Government Decision of the Republic of Moldova no. 626 of 20 August 2011.

These programs determine the objectives, the actions (measures), the performance indicators, the terms of implementation and the responsibility for the implementation.

Sources of Organic Matter Content and Nutrients in the Soil

There are various definitions and ways to calculate the efficiency of using nitrogen and its components. One of the definitions states that the efficiency of nitrogen use is equal to the ratio of grain yield per unit of available nitrogen in the soil, including the available nitrogen residues in the soil and nitrogen fertilizers. Nitrogen losses in the soil can be compensated for the following sources: atmospheric precipitations, plant debris, biological fixation, organic and mineral fertilizers. It has been established (Andrieş 2011) that once with the atmospheric precipitation in the soil, 6 kg N in the southern area and 9 kg/ha in the northern part of Moldova are introduced. According to estimates (Andrieş 2007), 5–6 kg/ha of biological nitrogen accumulates annually through the non-imbibition fixation mechanism (elementary nitrogen fixation by soil-free organisms) in the soil.

The symbiotic nitrogen fixation is, as is known, a process of atmospheric nitrogen fixation by nitrogen-fixing microorganisms, symbiotic with plants, especially from the Leguminous family (Lăcătuşu 2000). Moldovan research and university institutions (Programul complex de valorificare a terenurilor degradate şi sporirea fertilităţii solurilor. Partea II. 2004b) determined the amount of N biologically fixed from the atmosphere by these cultures. Alfalfa with a productivity of 30 t/ha green mass accumulates 150 kg/ha of nitrogen. In the soil, with vegetal remains and roots, 100–200 kg/ha of nitrogen remains. Soya fixes from the atmosphere 130 kg of nitrogen and peas is 80 kg/ha. According to recommendations (Lăcătuşu 2000, 2002, Lupaşcu 2004), the share of leguminous crops in crops should be 20–25%. Compliance with these recommendations provides 35–40 thousand tons of organic nitrogen in agriculture or about 20 kg/ha. Organic fertilizers are an important source of compensation for the loss of organic matter and nitrogen in the soil. Experimentally it has been established that for the preservation of organic matter and nitrogen in the soil, it is necessary to apply 10–15 t/ha of organic fertilizers (Андрианов et al. 1989, Андриеш et al. 1993, Боинчан 1999, Громов and Васильев 2007).

With this amount of organic fertilizer in the soil is incorporated 50–75 kg/ha of nitrogen. The achievement of this objective is only possible through the development of the zootechnical sector and the production of the respective quantities of organic fertilizers. Mineral fertilizers compensate for nitrogen deficiency in the soil are used to decompose fresh organogenic materials, synthesize humic substances and optimize the mineral nutrition of crop plants. Multi-annual agrochemical research has determined the necessity of Moldova's agriculture in mineral fertilizers

(Lupaşcu 1996, 2004). The average nitrogen dose for agricultural soils is 50 kg/ha. According to the latest estimates (Lupaşcu 2004), the share of biological nitrogen in agriculture will be around 60%, and the industrial nitrogen is 40%. The complex application of plant and agrotechnical measures will ensure the preservation of soil fertility, the achievement of expected crops and the protection of the environment of degradation and pollution.

Provision of Crops with Nitrogen

In the system of criteria characterizing the mode of nitrogen nutrition of plants in ontogenesis, agrochemical efficiency and ecological rationality of applied nitrogen fertilizers, an important indicator is the amount of nitrogen consumption, which is expressed in absolute values per unit area or product as well as the amount applied with the fertilizer. Every crop after harvest leaves for subsequent a certain amount of residues containing nitrogen. The latter enters the nitrogen pool of the soil and provides the course and direction of the processes of nitrogen transformation of fertilizers in the soil (Руделев 1974).

The combination of molecular nitrogen in the air and the replenishment of nitrogen in the soil is accomplished in two ways. A small amount of bound nitrogen (up to 3–5 kg per 1 ha) is formed in the atmosphere under the action of lightning discharges and in the form of nitric and nitrous acid enters the soil with precipitation.

The second way is the fixation of nitrogen in the air by nitrogen-fixing microorganisms living freely in the soil (azotobacter, clostridium, etc.) and nodule bacteria living in symbiosis with leguminous plants (biological synthesis of nitrogen) is more important for plant nutrition. Free-living nitrogen fixers assimilate up to 5–10 kg of nitrogen per 1 ha. The sizes of symbiotic nitrogen fixation depend on the type of legume. So, lupine can accumulate 100–170 kg of nitrogen, clover can 150–160 g, soybeans can 100 g, alfalfa can 250–300 g and vetch, peas and beans can 70–80 kg per 1 ha.

About 50%, bound by leguminous nitrogen, remains in root and crop residues, and after mineralization it can be used by crops, which follow in crop rotation after legumes. Thus, the total intake of nitrogen from the above sources far does not compensate for the removal of nitrogen by crops and losses from the soil as a result of denitrification and leaching. For this reason, in order to obtain high yields of any agricultural crops and improve the quality of the crop, the application of mineral nitrogen fertilizers obtained by artificial synthesis from air nitrogen (Муравин 2005) is of paramount importance.

The content of nitrogen and ash elements in plants and their organs can vary greatly and is caused by the biological characteristics of the culture, nutritional conditions and age. The amount of nitrogen in plants directly correlates with the protein content, and it is always more in seeds and young leaves than in the straw of the matured crops. There is less nitrogen in tubers and root crops than in leaves. Leafy vegetables (salad, spinach) have the highest ash content (up to 20% and more). The plants have significant differences in the composition and content of ash elements. Therefore, the knowledge about the macro and micronutrients for plants is needed. The functions of each of the macro and micronutrients in plants are strictly specific, not one element can be replaced by another. The lack of any of the macro or microelements causes a metabolic disorder and physiological functions in plants lead to a deterioration of their growth and development and a decrease in yield. With an acute shortage of even a single nutritional element, plants show characteristic signs of starvation and they die (Смирнов and Муравин 1977).

The nitrogen content in the soil should be optimal in relation to other nutrients. Doses of nitrogen fertilizers for the main crops: legumes— 30–40 kg/ha, annual herbs— 40–60 kg, winter cereals— 60–90 kg, spring grain— 30–60 kg, potatoes— 90–120 kg, root crops— 60–120 kg, fruit and berry— 60–90 kg/ha (Михайлова 2015).

The Effect of Nitrate-Nitrogen on Human Health

Today, there are questions about the study of the environmental consequences of the distribution of nitrates and nitrites in biological systems.

Soil organic matter is the main source of nitrates during the mineralization of which the nitrate stock of aquatic and terrestrial ecosystems is formed. Nitrogen-containing mineral fertilizers have become additional sources of nitrates; here you can also include wastewater from industrial enterprises, wastewater and waste from livestock breeding complexes, precipitation and municipal waste. All plants continuously extract nitrates from the soil and process them into organic compounds, such as amino acids, proteins and others.

However, nitrates are distributed in plants very unevenly. For the human body, nitrates themselves are harmless. But some of them turn into nitrites (nitrous acid salts), which can block cell respiration (Громов and Васильев 2007). Different plants have a different ability to accumulate nitrates. Leafy vegetables have the highest: cabbage, green cultures, salads and also root vegetables; less ar the eggplants, tomatoes and peppers. Nitrates are of particular concern due to their presence in the fresh waters of decentralized management (springs, wells).

Nitrates have a toxic effect on the human body. As a result of the systematic entry into the body, they can also have a carcinogenic effect. A high concentration of these substances affects the absorption of vitamin A and can lead to impaired work of the thyroid gland, heart and central nervous system. Nitrates, entering the human gastrointestinal tract, undergo numerous biochemical transformations. Under the action of microflora, they are restored to nitrite. The toxicity of the formed compounds is twenty times higher than the original. Once in the blood, nitrites interact with hemoglobin, changing the valence of iron and turn it into methemoglobin, which is no longer able to perform the function of a carrier of oxygen. For young children, the occurrence of methemoglobin in the blood is especially dangerous. Due to the low acidity in their stomachs, it contributes to the development of microorganisms that are involved in the conversion of nitrates to nitrites, the lack of well-formed enzyme systems for converting methemoglobin to hemoglobin and the consumption of large volumes of fluid per unit of body weight compared to an adult organism (Новиков et al. 1985). Recent calculations have shown that the nitrate-nitrite load for children aged from six months to six years is 84.2–111.2% more than in adults, even with the use of the same food. The weakening of the immune system and reduced mental and physical performance changes in the biocurrents of the brain and the appearance of persistent allergic reactions is a result of chronic intoxication of the human body with nitrates and nitrites. Methemoglobinia is not always accompanied by outwardly noticeable symptoms, which makes the diagnosis of this disease very difficult. Nitrites are included in metabolic processes, changing the activity of certain enzymes, increase directly or indirectly, the body's sensitivity to the effects of mutagenic and carcinogenic factors. Epidemiological studies have found a direct link between gastric cancer mortality and nitrate levels in food and drinking water (Андрианов et al. 1989).

For an adult, the maximum allowable rate of nitrates is 5 mg per 1 kg of a person's body weight, which means that a person weighing 60 kg has 0.25 g. For a child under six years old, the maximum permissible dose is 50 mg. And 10 mg of nitrates is enough for poisoning a baby.

The lethal dose of nitrates for a person is 8–15 g. It is relatively easy for a person to tolerate a daily dose of nitrates of 15 mg, 200 mg and 500 mg is the maximum tolerated dose (600 mg is already a toxic dose for an adult).

CONCLUSIONS

The soils of Moldova are characterized by relatively high total nitrogen content. The main amount of this nutrient is contained in the organic matter in the fractions of the non-hydrolyzed nitrogen (70–80%) and hardly hydrolyzed (10–15%).

These two fractions determine the structure and stability of the nitrogen in the soil. Mineral nitrogen content available to plants is only 1–2% of the total and insufficient value for plant nutrition and high yields of autumn wheat of 4.5–5.0 t/ha.

The amount of mineral nitrogen depends on the flow of organic matter and nitrogen into the soil. In the last 15 to 20 years, the flow of organic and nitrogen materials into the soil has declined considerably due to the lack of application of organic fertilizers in agriculture, the reduction of legumes in crops by five times and the reduction in mineral fertilizer (NPK) times.

The balance of organic matter and nutrients in agriculture is negative, and the effectiveness of nitrogen fertilizers is high.

In order to preserve soil fertility and obtain the expected yields, it is advisable to apply the following measures: agrochemical mapping of agricultural land and elaboration of the soil fertilization system for 5–6 years, periodic determination (over 4–5 years) of the balance of organic matter and nutrients in the soil, biological agriculture through the implementation of crop rotation, the application of 10–15 t/ha of organic fertilizers and the respective mineral fertilizer doses and the use of vegetable residues as organic fertilizer.

The population of the world is growing and in parallel increases the need for food. At the same time, the diet regimen is aimed at increasing meat consumption changes. Accordingly, an increase in meat production requires an increase in feed production. Since the basis of rations for feeding farm animals is grain, it is necessary to double world grain production by 2050 in order to meet the needs of a growing population and changes in dietary patterns. It is known that an increase in the production of plant products requires carrying mineral fertilizers. Over the past 50 years, the application of nitrogen fertilizers has increased by 20 times and, according to forecasts, its use will increase to 180 million tons by 2030. In addition to this, over the past decade, prices for nitrogen fertilizers have increased by more than 2.5 times.

However, not all of the nitrogen in the composition of plants comes from nitrogen fertilizers. Nitrogen utilization efficiency is a function of soil structure, climatic conditions, relationships between soil and bacterial processes, natural sources of organic and inorganic nitrogen, which are not included in the above formula. We recommend referring to the publication of Grahmann et al. (Grahmann et al. 2013) for a summary of new ways to determine the efficiency of nitrogen use.

There is also an obvious need for applied research on how to adapt nitrogen fertilizer management and equipment to prepare recommendations for farmers. Recommendations will be directed to specific areas and farms and will depend on abiotic factors, such as climate, soil type, rainfall and temperature as well as practicing agricultural practices, such as crop rotation, tillage intensity (zero, surface and minimum), amount of plant residues left in the field and finally the goals of the farmer to produce high yields of crops or to improve the quality of grain.

Acknowledgements

This work would not have been possible without the support and help of the Head of the Department of Environmental Chemistry, Director of the Institute for Agricultural and Forest Environment, Polish Academy of Sciences. I would like to express my very great appreciation to Prof. Dr. Lech Wojciech Szajdak for the valuable opportunity of this research work. His desire to give his generous support was highly appreciated.

I would also like to thank my colleagues from the Institute of Pedology, Agrochemistry and Soil Protection "Nicolae Dimo", who have allowed me to use the data of their research over the years of the nitrogen in the soils of the Republic of Moldova for this work research.

Finally, I wish to thank my parents and my family for their support and encouragement throughout my study.

REFERENCES

Andrieş, S., Ţîganoc, V. and Donos, A. 2000. Buletin de Monitoring Ecopedologic (agrochimie). Ediţia a VII-a. Chişinău, Pontos, p. 65.

Andrieş, S., Banaru, A., Cerbari, V., Demcenco, E., Donos, A., et al. 2001. Programul de conservare şi sporirea fertilităţii solurilor pentru anii 2011–2020 aprobat prin Hotărârea Guvernului Republicii Moldova nr. 626 din 20 August. p. 117.

Andrieş, S. 2007. Optimizarea Regimurilor Nutritive ale Solurilor şi Productivitatea Plantelor de Cultură. Chişinău, Pontos, p. 374.

Andrieş, S. 2011. Agrochimia elementelor nutritive. Fertilitatea şi ecologia solurilor. Chişinău, Pontos, p. 223.

Andrieş, S. et al. 2012. Recomandări privind aplicarea îngrăşămintelor pe diferite tipuri şi subtipuri de sol la culturile de câmp. Chişinău, Pontos, p. 68.

Andrieş, S., Cochină, V., Lungu, V. and Leah, N. 2016. Măsuri şi procedee agrochimice de sporire a fertilităţii solurilor erodate. Simpozion internaţional ştiinţifico-practic "Utilizarea eficientă a resurselor hidro-funciare în condiţiile actuale. Realizări şi perspective". Chişinău, pp. 51–54.

Andrieş, S. 2017. Materia organică în solurile Moldovei şi măsuri de sporire a fertilităţii. Akademos, 2: 71–77.

Anuarul Statistic al Moldovei. 2012. Chişinău, pp. 210–216.

Banaru, A. 2002. Îndrumări Metodice Perfecţionate Pentru Determinarea Bilanţului de Humus în Solurile Arabile. Chişinău, p. 22.

Burlacu, I. 2000. Deservirea agrochimică a agriculturii în Republica Moldova. Chişinău, Pontos, p. 228.

Cadastru Funciar al Republicii Moldova. 2012. Chişinău, p. 985.

Camberado, J. 2012. Bioavailability of Nitrogen. Handbook of Soil Sciences. Resources, Management and Environmental Impact, 2nd Ed. Chişinău, Pontos, pp. 11.1–11.14.

Cerbari, V. 2010. Monitoringul Calităţii Solurilor Republicii Moldova (Banca de date, concluzii, prognoze, recomandări). Chişinău, Pontos, pp. 475.

Donos, A. 1994a. Problemele azotului în agricultură şi căile dezvoltării lor. Revista Agricultura Moldovei. 12: 12–14.

Donos, A. 1994b. Bilanţul de azot în agricultură. Revista Agricultura Moldovei. 11–12: 7–8.

Donos, A. 1994c. Determinarea bilanţului substanţelor nutritive în sol. In: Recomandări Privind Aplicarea Îngrăşămintelor. Chişinău, Agroinformreclama, pp. 134–138.

Donos, A. 1995. Evoluţia ciclurilor de transformare a azotului pe terenurile arabile din Republica Moldova. *In*: Evaluarea naturalistică şi economică a resurselor de sol. Iaşi, 9.

Donos, A. 1997. Circuitul azotului în agricultura Moldovei şi măsurile generale de menţinere a unei evoluţii de bilanţ echilibrat pe solurile arabile. *In*: Problemele agrochimiei în agricultura contemporană. Chişinău, pp. 106–108.

Donos, A. 1998a. Bilanţul humusului şi elementelor biofile în solurile erodate din Moldova. In: Resursele funciare şi acvatice. Valorificarea superioară şi protecţia lor. Vol. II. Chişinău, pp. 52–63.

Donos, A. 1998b. Postacţiunea îngrăşămintelor de recoltă a grâului de toamnă şi porumbului pe cernoziomul obişnuit. *In*: Resursele funciare şi acvatice. Valorificarea superioară şi protecţia lor. Vol. II. Chişinău, pp. 64–70.

Donos, A. 2001. Ciclurile de transformare şi acumulare a azotului molecular în sol. *In*: Solul şi Viitorul. Chişinău, pp. 207–208.

Donos, A. and Andrieş, S. 2001. Instrucţiuni metodice perfecţionate pentru determinarea şi reglarea bilanţului de elemente biofile în solurile Moldovei. Chişinău, Pontos, p. 24.

Donos, A. 2006. Căile de acumulare a azotului în solurile arabile. IEFS, Chişinău, p. 28.

Donos, A. 2008. Acumularea şi transformarea azotului în sol. Chişinău, Pontos, p. 206.

Grahmann, K., Verhulst, N., Buerkert, A., Ortiz-Monasterio, I., Govaerts, B. 2013. Nitrogen use efficiency and optimization of nitrogen fertilization in conservation agriculture. CAB Reviews 8, No. 053, http://www.cabi.org/cabreviews. Online ISSN 1749-8848.

Instrucţiuni metodice perfecţionate pentru determinarea şi reglarea bilanţului elementelor nutritive în solurile Moldovei. 2001. Chişinău, pp. 24.

Lăcătuşu, R. 2000. Mineralogia şi Agrochimia Aolului. Iaşi, p. 252.

Lăcătuşu, R. 2002. Dicţionar de Agrochimie. Bucureşti, p. 312.

Lupaşcu, M. 1996. Agricultura Moldovei şi Ameliorarea ei Acologică. Chişinău, Ştiinţa, p. 112.

Lupaşcu, M. 2004. Lucerna: Importanţa Ecologică şi Furajeră. Chişinău, p. 304.

Nour, D. 2004. Eroziunea Solului. Chişinău, Pontos, p. 476.

Program complex de valorificare a terenurilor degradate şi sporirea fertilităţii solurilor. 2004. Partea I. Ameliorarea solurilor. Chişinău, Pontos, p. 212.

Programul complex de valorificare a terenurilor degradate şi sporirea fertilităţii solurilor. Sporirea fertilităţii solurilor. 2004. Partea II. Chişinău, Pontos, p. 135.

Recomandări privind aplicarea îngrăşămintelor. 1994. Chişinău, p. 169.

Rusu, A. 2008. Îngrăşămintele organice şi resturile vegetale ca factori de majorare a rezistenţei solului la fenomene ecologice nefavorabile. În: Diminuarea impactului factorilor pedoclimatici extremali asupra plantelor de cultură. Chişinău, pp. 78–92.

Tan, K.M. 2000. Environmental Soil Science, 2nd Ed. New York, pp. 80–144.

Tate, R. 2000. Soil Microbiology. 508.

Ursu, A. 2011. Solurile Moldovei. Chişinău, Ştiinţa, p. 321.

Zagorcea, C. 1989. Evoluţia circuitului şi bilanţului elementelor biofile în agrofitocenozele din Republica Moldova în ultimul secol. Resursele funciare şi acvatice. Valorificarea superioară şi protecţia lor, Vol. 2. Chişinău, pp. 121–125.

Агроклиматические ресурсы Молдавской ССР. 1982. Ленинград: Гидрометеоиздат, p. 198.

Андрианов, А.П., Ильнцкий, А.П., Славная, И.Л. and Стрижак, К.К. 1989. Онкологическая профилактика—путь к снижению онкологической заболеваемости. М. p. 76.

Андриеш, С.В., Донос, А.И. and Лях, Н.М. 1993. Профильное распределение нитратов в почвогрунтах при систематическом применении удобрений. Известия АН Республики Молдова, сер. биол. и хим. наук, 40: 67–70.

Андриеш, С.В. 1993. Регулирование питательных режимов почв под планируемый урожай озимой пшеницы и кукурузы. Кишинев, Штиинца, p. 200.

Алексеев, В.Е. 1999. Минералогия Почвообразования в лесостепной и степной зонах Молдавии. Кишинев, p. 87.

Боинчан, Б.П. 1999. Экологическое земледелие в Республике Молдова. Кишинев, Штиинца, p. 268.

Донос, А.И. 1977a. Формы азота в серой лесной почве и их изменение при интенсивном сельскохозяйственном использовании. Димовские чтение. 11–12. Кишинев, Штиинца. 48–49.

Донос, А.И. 1977b. Изменение фракционного состава азота при систематическом применении минеральных удобрений. Тез. Докл. Совещ. участников Географ. сети опытов с удобрениями Молдавской ССР и Украинской ССР. Ч. II. Кишинев, Штиинца. 152–153.

Донос, А.И. 1978a. Содержание форм азота в выщелоченном черноземе при систематическом применение минеральных удобрений. Ч. II. Кишинев, Штиинца. 138–139.

Донос, А.И. 1978b. Влияние минеральных удобрений на содержание аминокислот органических фракций азота в выщелоченном черноземе. Кишинев, Штиинца. 100–107.

Донос, А.И. 1978c. Изменения азотистых соединений в обыкновенном черноземе при интенсивном сельскохозяйственном использовании. Агрохимия, 7: 91–97.

Донос, А.И. and Кордуняну, П. 1978. Содержание и динамика аминокислот органических фракций азота в обыкновенном черноземе. Почвоведение, 9: 46–53.

Донос, А.И. 1979. Формы азота в почвах Молдавии и их изменение при систематическом применение минеральных удобрений. М. p. 17.

Думитрашко, М.И. 1987. Пути повышения плодородия черноземов и продуктивности полевых севооборотов при различных системах удобрения. Автореф. дисс. д.с.-х.н. Минск. p. 35.

Громов, В.И. and Васильев, Г.А. 2007. Здоровая жизнь. Сборник. М. p. 174.

Крупеников, И.А. and Подымов, Б.П. 1987. Классификация и систематический список почв Молдавии. Кишинев, Штиинца. p. 157.

Кордуняну, П., Пресман, В. and Донос, А. 1997. О накоплении азота в почве. Сельское хозяйство Молдавии. 4: 31–33.

Крупеников, И.А. 2008. Черноземы. Возникновение, совершенство, трагедия деградации, пути охраны и возрождения. Chişinău, Pontos, p. 285.

Крупеников, И.А. 1967. Черноземы Молдавии. Кишинев: Картя Молдовеняскэ, p. 427.

Крупеников, И.А. 1992. Почвенный покров Молдовы. Прошлое, настоящее, управление, прогноз. Кишинев, Штиинца, p. 263.

Загорча, К.Л. 1990. Оптимизация системы удобрения в полевых севооборотах. Chişinău, Ştiinţa, pp. 288.

Ласе, Г.А. 1978. Климат Молдавской ССР. Л. p. 378.

Михайлова, Л.А. 2015. Агрохимия: Курс лекций – часть 1, Удобрения: виды, свойства, химический состав//Пермь, ИПЦ «Прокрость», p. 427.

Муравин, Э.А. 2005. Практикум по агрохимии: практическое пособие. М., Колос, p. 288.

Новиков, Ю.В., Окладников, Н.И., Сатдиутдинов, М.М. and Андреев, И.А. 1985. Влияние нитратов и нитритов на состояние здоровья населения // Гиг. И сан. 8: 58–62.

Нормативы по использованию минеральных и органических удобрений в сельском хозяйстве Молдавской ССР. 1987. Кишинев, p. 37.

Пономарева, В.В. and Николаева, Т.А. 1980. Содержание и состав гумуса в черноземах Стрелецкой степи под различными угодьями//Тр. Центрально-Черноземного заповедника. Вып. 8.: 209–235

Почвы Молдавии. 1984. Т.1. Кишинев, Штиинца, p. 352.

Почвы Молдавии. 1986. Т.3. Кишинев, Штиинца, p. 336.

Руделев, Е.В. 1974. Превращение азотных удобрений в почве и использование их растениями в зависимости от биологических особенностей сельскохозяйственных культур//Бюллетень ВИУА, М., 22: 30–34.

Синкевич, З.А. 1989. Современные процессы в черноземах Молдавии. Кишинев, Штиинца, 176–181.

Смирнов, П.М. and Муравин, Э.Л. 1977. Учебники и учебные пособия для высших учебных заведений/Агрохимия/Москва «Колос», p. 304.

Смирнов, П.М. 1977. Проблемы азота в земледелие и результаты исследований с N. Агрохимия. 1: 33–47.

Смирнов, П.М. 1979. Газообразные потери азота почв и удобрений и пути их снижения. М., Наука, pp. 56–65.

Турчин, Ф.В. 1960. Хроматографический метод определения аминокислотного состава азотных органических веществ в почве. М., АНСССР, pp. 235–240.

Шконде, Е.И., Королева, И.Е. 1964. О природе и подвижности почвенного азота. Агрохимия, 10: 17–35.

Шконде, Е.И. 1965. Формы азота в почве и их определение. Киев, pp. 28–33.

Щербаков, А.П. 1968. Формы азота и их изменение при систематическом использовании в почвах ЦЧО. Воронеж, p. 18.

Chapter 6

Composting and Quality Improvement Processes for Active Protection of the Environment

Paula Ioana Moraru* and Teodor Rusu

University of Agricultural Sciences and Veterinary Medicine Cluj-Napoca, Romania.

INTRODUCTION

Composting is the technique of transforming organic waste into fertilizers, comprising all microbial, biochemical and physical transformations that vegetal or animal waste suffer from in their initial state and up to different stages of humification.

Among the waste sources used today in composting, one can remember agriculture, industrial waste and last but not least from the local administrations (parks, beaches and wastewater treatment plants). The biggest problems related to huge amounts of waste are household waste and animal waste.

The production of organic waste is a continuous growing process, which results from products based on the loss of organic matter from the soil due to intensive agriculture and poor management of the soil, urbanization and climate conditions. This situation imposes the development of recycling processes of organic waste as an alternative to other procedures, like incineration or storing of waste.

Composting can be done at different levels of human activity, from the gardener who produces his own 'black gold' (compost) to industrial waste composting facilities.

---

*For Correspondence: E-mail: moraru_paulaioana@yahoo.com

If the gardener applies simple rules to composting, without being concerned with the microbiological or chemical process, the execution of composting operations at a city and commercial level requires knowledge of all physical-chemical and biological parameters that ensure fast waste processing at sufficiently high temperatures, which allows obtaining an odorless compost and pathogenic microorganisms.

In composting, we can practically use all organic waste that is produced in the environment (Solti et al. 2006):

- Organic livestock: manure (beef, horse, pig, poultry, sheep), slurry, liquid fertilizer.
- Organic vegetable matter: vegetable crop residues (straw, corn cobs), wood waste, sawdust and plants, tree bark, grass, foliage, etc.
- By-products from the textile, leather, paper, food (canning industry, spirits industry, etc.).
- Household waste (kitchen and garden waste, paper, residual sludge, etc., collected selectively).

In order to obtain a good quality compost, we need to know the properties of the materials used and the composting conditions so that we can effectively manage the composting process.

WASTE USED IN COMPOSTING MANURE OBTAINED IN ZOOTECHNICS

Organic waste coming from the zootechnics activity raises serious problems in terms of storing space and biodegradation processes. This process, on the one hand, develops germs and a bad smell, but on the other hand, it can ensure the conversion of waste into biofertilizers or biofuels.

The manure obtained in zootechnics is an important source to obtain good compost. In general, manure from zootechnics has two forms: the liquid one, also called decanted wastewater, and the solid form of sludge, respectively, called manure. The problems waste create are multiple among which the high volume of decanted wastewater and sludge and the pollution of surface water when the evacuation of wastewater is done in streams close to zootechnical facilities, big surfaces for storing solid manure, etc.

In Romania, once big livestock farms appeared in the last decades of the last century and also wastewater treatment plants, the need to value sludges in agriculture as organic fertilizers appeared too. Sludge from biological lakes or lagoons, after water is pumped (which diluted, it can be used at crop irrigations), is left to dry. After being dry, it is transported to the lands which are going to be chemically fertilized. This is a form of direct value of them. The application of sludge is regulated by the Order of the Ministry of Environment and Water Management no. 344 from the August 16, 2004, to approve the technical Norms regarding the protection of the environment and especially of soils when sewage sludge is used in agriculture.

Special issues came up related to the reduction of wastewater treatment plants of industrial facilities to raise pigs with high amounts of partially dehydrated sludge (minimum 70% humidity). Total dehydration (which solves the problem from a hygienic-sanitary point of view) is too expensive and supposes much energy to be worth it. The content of this sludge is rich in nutritious elements for plants, related in organic combinations easily biodegradable. An easy calculation shows that by enriching the soil by 10 tons of sludge for a hectare, 45.3–102 kg/ha N, 28.5–60 kg/ha P_2O_5 and 3.3–9 kg/ha K_2O are administered. Such big variations of contents of nutritious elements are due to the quantity and quality of food given to animals within a year.

The contents of heavy metals in the soils on which sludge is applied and the maximum annual amounts of these heavy metals which can be introduced in the soils used in agriculture are presented in Tables 6.1–6.3 (Order 344/2004).

Table 6.1 Maximum values allowed for the concentrations of heavy metals in the soils on which sludge is applied (mg/kg of dry matter in a representative sample of soil with a pH higher than 6.5)

Parametres	*Limit Values*
Cadmium	3
Copper	100
Nickel	50
Lead	50
Zinc	300
Mercury	1
Chrome	100

Table 6.2 Maximum concentrations of heavy metals allowed from the sludge meant to be used in agriculture (mg/kg of dry matter)

Parametres	*Limit Values*
Cadmium	10
Copper	500
Nickel	100
Lead	300
Zinc	2,000
Mercury	5
Chrome	500
Cobalt	50
Arsenic	10
AOX (sum of organohalogenated compounds)	500
PAH (Polycyclic aromatic hydrocarbs)	5
PCB (Polychlorinated biphenyls)	0.8

By knowing the value of animal manure for agricultural production, it needs to reach the lands, to be administered and incorporated in the soil. But due to the seasonal character of soil tillage and the small amounts of manure that accumulate daily, it needs to be stored.

Table 6.3 Limit values for the annual amounts of heavy metals which can be introduced in agricultural lands based on an average of 10 years (kg/ha/year)

Parametres	*Limit Values*
Cadmium	0.15
Copper	12
Nickel	3
Lead	15
Zinc	30
Mercury	0.1
Chrome	12

According to the type and duration of storing, there are two technologies of biotransforming: extensive composting and intensive composting.

Extensive composting represents the storing of manure in big piles where the transformation of vegetal waste and animal manure produces regularly in the absence of air, in the presence of too much water and under conditions of strong compaction. Under these conditions, decomposing occurs slowly, incompletely and at temperatures that do not exceed 40°C even during summer. In this case, only the upper layer of the pile shall be fermented. The lower layers of the pile shall be incompletely fermented, recording thus the accumulation in big quantities of organic acids, some of them are toxic for plant roots and the weed seeds and pathogens shall not be destroyed.

Intensive composting can be done at all collecting levels from the small domestic system up to industrial facilities, if certain rules of ferment and decomposing, which lead to obtaining a high quality composting are met.

According to the setting technology of the composting pile, the composting history states three types of composting:

- Intensive anaerobic composting.
- Intensive anaerobic composting with an aerobic layer followed by a prolonged phase in anaerobiosis.
- Intensive aerobic composting.

One of the big disadvantages of applying manure, frequently met in all zootechnical farms is storing the manure at the edge of the fields or directly on the agricultural lands, resulting in a continuous loss of nitrogen, energetic and food carbon and last but not least the pollution of the environment.

The Common Agricultural Policy imposes through Ecoconditionality for each farm the solving of manure problem from zootechnics, and composting is one of the simple, practical methods with a lot of beneficial results upon the soil fertility; a method which can be applied both at the level of big livestock farmers and by small farmers.

In order to protect all water resources against pollution with nitrates coming from agricultural sources, according to the Code of good agricultural practices, farmers must (Order 999/2016):

a) Have storing capacities of manure without structure defects, the size of which must exceed the storing need of manure and take into account the longest period of prohibition for applying organic fertilizers on the agricultural land; storing the manure can be made in permanent deposits (individual and/or communal system) or temporary deposits in the field, on the land where it will be spread.

b) Meet the prohibition period for applying manure on the agricultural land (Table 6.4).

c) Not exceed the quantity of 170 kg nitrogen/ha coming from applying organic fertilizers on the agricultural land during a year; in this regard, they must follow a fertilization plan by meeting the standards regarding the maximum quantities of nitrogen fertilizers which can be applied on the agricultural land and ensure a uniform distribution of fertilizers on the agricultural land.

d) In the case of exploitations which practice agriculture in the irrigated system or in which the planned production needs higher quantities of nitrogen than the ones stated by the standards regarding the maximum quantities of nitrogen fertilizers, which can be applied on agricultural land to follow a fertilization plan set up based on an agrochemical study.

e) Not apply organic or mineral fertilizers during rain, snow and strong sun or on lands full of water, covered by snow or if the soil is strongly frozen/deeply cracked/dug in order to install some drainage/subsoiling works; on soils saturated with water, flooded, frozen or covered by snow, it is forbidden to apply organic animal fertilizers, thus avoiding losses of nitric nitrogen with percolation water and leakage as well as losses by denitrification as elementary nitrogen or nitric oxides.

f) Ensure the incorporation in the soil of organic fertilizers applied on arable lands with a slope higher than 12% in 24 hours at the most from their application.

g) Not apply organic or mineral fertilizers on protection strips existing on agricultural lands situated next to protected areas of surface water or on agricultural lands situated in protected areas of surface water or sanitary and hydrogeological protection areas of capture sources of drinking/mineral water and therapeutic lakes; the minimum width of protection strips is 1 m on lands with slope up to 12% and 3 m on lands with a slope higher than 12%, the land slope being the average slope of the physical block adjacent to the water stream.

h) To complete daily, keep for 5 years and present for control the evidence documents of the exploitation regarding the agricultural surface used, crop structure, number of livestock, type and quantity of nitrogen fertilizers applied on the agricultural land (fertilization plan) and the fertilizer delivery/shipping documents.

The prohibition period for applying manure on the agricultural land is defined by the period when the average air temperature drops below 5°C, and the crop demands compared to nutrients are reduced and the risk of percolation/surface leakage of nitrogen is high. Manure is regularly administered during autumn at the basic soil tillage (plowing with a return of the furrow), under bad weather conditions, especially when it is cloudy and slowly windy.

Table 6.4 Calendar of prohibition for applying organic fertilizers (CBPA, 2015)

<table>
<tr><th>Specification</th><th></th><th></th><th>Period of prohibition</th></tr>
<tr><td>Solid Organic Fertilizers</td><td colspan="2">Arable Land and Pastures</td><td>1 November–15 March</td></tr>
<tr><td rowspan="3">Liquid Organic Fertilizers</td><td rowspan="2">Arable Land</td><td>Autumn Crops</td><td>1 November–1 March</td></tr>
<tr><td>Other Crops</td><td rowspan="2">1 October–15 March</td></tr>
<tr><td colspan="2">Pastures</td></tr>
</table>

COMPOSTING OF THE MANURE

Daily fresh manure from the stable cannot be used directly as its daily application is impossible. Therefore, it is mandatory to keep it, ferment it and store it for a shorter or longer period. The purpose of ferment (maturation) is the reduction of the C:N ratio, and the nutritious elements should become more accessible. The ferment is very important in obtaining manure with a rich humidifier yield. Upon its incorporation in the soil, when an organic fertilizer has an advanced degree of decomposing in the platform, it ensures a humus yield of 30–50% of the dry substance, a superior humus yield compared to the direct incorporation of straws or fresh manure. The final humification in the soil shall be done in better conditions if the straws and manure are decomposed, transforming into a loose mass which can be administered uniformly.

The storing is thus motivated based on biological reasons but also on usage, exploitation and handling reasons of the manure. The structure of the vegetal production does not allow taking out and daily incorporating the manure, one needs to wait for the harvest of prior crops from the lands which are going to be fertilized in order to be incorporated in the soil.

Manure is fermented under the influence of microorganisms, which in the process of decomposing of organic substances also runs a synthesis activity, which precedes the process of humification that started in the platform and continues in the soil. 'The reorganization' of nitrogen and the other nutritious elements, their transformation into more accessible forms, together with the decomposing of the straw layer form an excellent nutritious sublayer for microorganisms, which grow rapidly in number (Amlinger et al. 2003).

During the ferment of the manure there are two phases:

During the oxidation phase, in the manure mixed with straws, the temperature reaches quickly 50–70ºC. During decomposing, water and carbon dioxide are formed from non-nitrogen substances and is considered as normal losses from the manure mass. The nitrogenous substances from manure are transformed under the influence of the urease enzyme into ammonium carbonate, which cedes ammonia. The process is known by the name of ammonification and is done by slow microbiological decomposing by forming NH_3, CO_2, H_2O, H_2S, etc. Following the oxidation of the ammonia and the nitrification process, it is transformed into nitrous nitrogen and then into nitric nitrogen (process done with heat release), which decomposes through denitrification, resulting in free nitrogen that evaporates. In order to avoid nitrogen loss, the phase of aerobic oxidation cannot exceed 3–5 days. That is why after 3–5 days from the settlement of a manure layer, the oxidation phase must end by adding a new layer of fresh manure or soil.

In the denitrification phase, the activity of microorganisms is reduced due to the lack of oxygen. Through denitrification, nitrates are replaced by oxygen, which contributes to the oxidation of the organic matter and CO_2 is released. The carbon dioxide forms together with ammonia the ammonium carbonate, which has a slower decomposing. In this phase, manure ferments on an average 100 days, reaching a darker color, the material becomes more homogeneous and the C:N ratio reaches the desired value of almost 20:1. Fermenting will be more profitable if humidity oscillates around 25%. During dry summer, water soaking is recommended.

The assessment of the quality of fermenting can be made according to the appearance of the manure settlement. In the case of half fermented manure, straws can still be easily distinguished and have a lighter color. In the case of well-fermented manure, straws can barely be seen, have a darker color and texture is more homogenous. Such manure is the most valuable from a biological point of view and is easy to administer because it spreads easily. Overmature manure is poorer in nutritious substances and hard to be spread.

During fermenting, the weight loss is 20–25%, the loss by fermenting is around 25% and it can reach even 50%. Losses are higher in nitrogen, but they can also be in phosphorus and potassium (Table 6.5).

Table 6.5 Content of manure according to quality (kg/t)

Content of Nutritious Substances	*Good*	*Average*	*Weak*
Nitrogen	5.08–8	4–5	3–4
Phosphorus (P_2O_5)	2.5–5	2–2.5	1.5–2
Potassium (K_2O)	6–8	5–6	3–5
NPK	13–21	11–13.5	7.5–11
Organic matter	18–22	15–18	10–15
C:N ratio	15–20:1	20–25:1	25–30:1

Fermenting Methods

Fresh manure can be fermented through several methods among which the most well-known are the following:

1. Spreading the manure on the platform: In this, the manure is taken out daily from the stable and is spread uniformly on the manure platform. Thus, the manure makes contact with the air on a large surface; during summertime, it dries fast and during wintertime, it does not get warm to the necessary temperature. Therefore, it ferments with big losses for a long time and unevenly. It can be considered as the most inadequate composting procedure of manure which has to be avoided.
2. Sequential fermenting of manure: Manure is fermented in so-called manure piles, 4 m wide, generally 20–25 m long, which takes place gradually. After a covering of up to 3 m high, during the three-month ferment, it drops to 2.5 m. The manure pile is set on a concrete foundation, equipped

with a urine basin on the bottom of which in order to absorb the urine, one shall set straws, chaff, chopped straws or turf in layers 25–30 cm thick. The manure taken out daily must be spread on all the pile's width and at a thickness of 50–60 cm. The manure stays like this for 2–3 days during the summertime and 3–5 days during the wintertime to trigger the oxidation process. Within the next few days, the manure taken out is set near the manure from the first day, and over the section from the first day, one shall put another layer only after 2–3, respectively 3–5 days. The operation shall continue until the section started shall be 3 m high on the length of 20–25 m. Upon forming the manure pile, one must be careful that the sidewalls should be abrupt.

3. Manure can be fermented in manure platforms, in the paddock and stableyards: The manure platform is the place on which the manure pile is set. This can have a permanent character near stables or a temporary character in the corner of the batches which are going to be fertilized. On the batches which are going to be fertilized, it is appropriate to have the parallel settlement of the manure piles at opposite corners; thus during applying, the manure can be spread and incorporated as soon as possible. The parallel settlement eases the mechanized loading, the loading machines can enter among the manure piles, loading to the right and also to the left.

   Placing the manure platform near the batches has the advantage that the fermented manure reaches the soil more quickly and with fewer losses of nutritious substances. Therefore, the household shall be cleaner, and the pollution of the environment more reduced if on its field the ferment and storing of big quantities of manure do not take place. Even in this case, it is necessary to form a small manure platform in the household, where the daily quantities of manure can be evacuated and fermented.
4. The best and oldest method to ferment manure is the one which takes place in stableyards. Under free animals, which are able to move freely (foals, young taurine and sheep), one shall set a lot of bedding, and the manure shall not be taken out daily; it can be taken only after 3–4 months after the organic matter is already fermented. Rich bedding is absolutely necessary as it not only keeps the animals clean but also for the total intake of urine. As sheep have relatively little urine, the surface of the manure during summertime must be watered with water, then covered with dry bedding so that the sheep fur does not moisten. The manure treated in such a way it remains humid, compact and ferments well all the time. The advantage of the method of fermenting manure in the yard is that fluid manure absorbs without losses, and the manure becomes richer in nutritious substances. For the bedding one can use besides straws other materials too, such as rapeseed strains and corn cobs which following people's stepping on them and permanent moisture decomposing it easily. The manure ferments under the roof; it is neither exposed to the drying effect of the wind and the sun nor it is leached by rain. The method makes it possible to give up building urine tanks as a smaller workforce is needed.

Yard manure has a nitrogen content nearly 50% higher than the manure fermented in manure piles. The disadvantage is that in free air this high content of nitrogen decomposes more easily, so it is important that yard manure should reach the batch as quickly as possible and be returned under the furrow on the same day.

As the best quality manure is the yard one, and the fermenting method is the cheapest and with the smallest demand of the workforce; it is advisable to build adequate facilities for maintenance in the yard. The maintenance in the yard is part of the systems recommended to raise livestock under ecological conditions.

The ferment in the yard of the sheep manure, through the share it has in ecological agriculture in Romania, can be an important source for soil fertilizing. Its ferment has certain features as it is considered dry, warm manure and richer in potassium compared to the other types of manure (Table 6.6). A sheep of 45 kg produces around 5 liters of manure per day, of which 80% feces and 20% urine.

Table 6.6 Content in nutritious substances of sheep manure obtained in the yard

Content of Nutritious Substances	%	g/liter
Organic matter	28	250–350
N total	0.5–0.7	1.5–4.0
P_2O_5	0.3–0.4	0.5–2.0
K_2O	0.9–1.2	1.5–3.0
CaO	–	0.54
MgO	–	0.16

Adjusting the C:N ratio to ensure ferment is made through bedding of 20 kg straws/day for 60–80 sheep. To adjust humidity, one shall add humid organic materials or water.

If it is necessary to transport fermented manure in the yard next to the fields which are going to be fertilized and the field is not free, its storing is done in big piles. Their dimension resembles one of the manure piles, that is 4 m wide, 20–25 m long and 3 m high. Upon its settlement, one shall take into account similar criteria, that is it must be placed at the opposite corners of the batch in a parallel position so that the middle of the pile should be up and the land covering should be convex and making it possible for rain to fall. The walls must be modeled so that they become steep and the surroundings dug, thus preventing rain from entering deep.

COMPOSTING OF URBAN HOUSEHOLD WASTE

Industrial composting came up as a need when municipalities could not deal with the huge amount of domestic garbage that polluted the environment and the outskirts of cities and when the laws against those who polluted by randomly burning the garbage were tightened.

Even higher problems were raised and are still raising the sludges separated from town wastewater. Their burning, just like domestic garbage, specially built ovens, becomes more and more expensive once with the increase of oil price, also leading to the pollution of the atmosphere.

At the same time, the waste composition is very diverse and its management must take this into consideration. For example, the average composition of domestic garbage in Braşov, in 2002, out of the total collected quantity of 290,500 m^3 (80000 t) was carton 9%, glass 3%, metal 8%, plastic 7%, textile 1%, organic materials 65–70% and others 2–7% (Dumitrescu 2004).

The collection, handling, recycling and storing of urban domestic waste becomes a more and more acute problem for all humankind. Several methods of processing urban solid and organic waste have been developed and are still under improvement, like:

- Storing in ecologically protected pits;
- Incinerating with or without production of thermal or electric energy;
- Composting;
- Obtaining biogas;
- Ecological pits with a collection of biogas;
- Recycling reusable materials (metal, glass and paper), etc.

Ecologically protected pits are storing spaces for domestic garbage, isolated in the lower part against infiltrations to the groundwater, by successive layers of clay and isolating foils; the leaks of liquid materials being collected in the lower part of the pits through special collectors.

The liquid collected is either reused for re-wetting or sent to water treatment plants. In the upper part, the insulation is also made by layers of earth and clay. In the surrounding area of the ecological pits, there are wells drilled to control the infiltrations in the groundwater and the emitting of methane gas. Certain improved ecological pits can be equipped with installations for capturing methane gas coming from anaerobic processes; this gas is being used either directly as for fuel or to generate steam, respectively creating electrical energy.

As for incinerators, the main constructive types developed up to now are with fixed fireplace, rotary tubing, in a fluidized bed and steps. The main characteristic of all the incinerators is making the combustion of waste in minimum two steps: a primary step of gasification of solid waste and partial burning of heavy combustible components and a secondary step for burning combustible gas and easily engaged components. The burning is followed by recovering the heat with steam generation and purification of burned gas in several steps.

In order to neutralize hospital and animal waste, incineration is almost the only solution. The incinerating installations for such waste must meet special temperature conditions (substantially higher than those from the incinerators for non-dangerous waste) and time to maintain waste at a high temperature in order to destroy all pathogenic germs.

The main disadvantage of incinerators is the danger of emissions in the atmosphere, once with burned gas, of certain extremely toxic components like dioxin and furans, appeared by burning under certain conditions of waste which contains halogenated hydrocarbons (especially PVC).

Generating biogas is done either in horizontal basins, sealed at the surface with foils or tarpaulins from plastic materials, or in vertical cylindric tanks equipped with agitators. The principle of forming biogas (containing mainly methane gas) is the anaerobic ferment of organic materials (especially biological debris from zootechnical farms). Solid debris stabilized after ferment can be used or valued as natural fertilizers. There are also domestic biogas installations, dimensioned to take over organic waste from a household, which ensures the fuel gas is needed for cooking, even to heat the house or for certain zootechnical spaces.

Composting is a procedure of recovering and biological decomposing of organic materials from waste as it is finally processed as a granulated or chopped product, used as a vegetal layer in agriculture, horticulture, etc. As compost, one can recover almost 50% of the volume of domestic waste, and composting can be considered actually one of the natural recycling components of nutritious elements necessary for the animal and plant life.

The main three technologies for storing urban domestic debris (recycling, incinerating and storing in ecologically protected warehouses) are now used at the same time in most of the world, even if their use is done in substantially different proportions from country to country.

Waste integrated management is at the basis of all strategies and recovery plans, having as a fundamental element the differential collection.

If the Prefect of Paris, POUBELLE, guessed at the end of the nineteenth century, the need of the garbage collector, which carries his name ever since not in every case, society understands today the meaning of actions of a differential collection of waste.

The steps which have to be made are reducing the production of waste, recovery, reuse and recycling, incinerating with energy recovery, incinerating hospital waste and toxic and dangerous products and composting in order to obtain ecological fertilizers and storing in installations according to norms.

The obligation of recovery or removing waste is mainly of the waste producer. If there is not a sanitation system implemented by somebody else, the waste producer must do the necessary activities, thus:

- He must use recycling devices.
- Uncontrolled storing is forbidden.
- In rural areas, recovery at the scene must be ensured by composting waste from households.

All the policies regarding waste meet at their basis the following principles:

- The prevention principle: preserve natural resources following the reduction of amounts of waste generated.
- The precaution principle: reducing the impact upon the environment and human health.
- The 'producer pays' principle: emphasis on the responsibility of waste producers.
- The proximity and self-sufficiency principle: making an integrated and adequate network of installations and facilities to eliminate waste.

COMPOSTING OF DOMESTIC WASTE

Currently, there is a special interest in using biological processes to solve certain problems related to environment protection, like warehouses of solid waste, dangerous waste and treatment of polluted water. One of the biological processes which imposed for the remedy of organic solid waste is the controlled aerobic decomposing, called composting, applied at all administrative and social levels (Ruggieri et al. 2009).

The selective collecting of waste from your household and obtaining compost in your garden is one of the most important measures each person can take and be aware of the impact of his/her daily activity upon the environment. The compost pile from the garden should be a very usual thing. Fruit and vegetable scraps, eggshells, coffee grounds, flowers, leaves, grass, etc.; in other words, all organic residues from the household can be used for composting. The practices developed in the rural area during the last decades through which in springtime the household waste (vegetal debris, leaves, grass, rubbish, branches, etc.) are gathered and burned are not the best solution for their valuing. Out of organic debris, one can obtain by forming compost, precious humus, destined to improve the soil quality and at the same time to reduce the amount of waste taken to a trash can.

The composting of domestic waste can be done in piles or special containers for compost, bio-trash cans. This depends mainly on the available space from the garden. For the compost piles, a surface of at least 1.50 × 1 m is needed, while for a container—in dedicated stores there are several types—less room is needed. It is important that the container for compost should not be hermetically sealed because thus the air cannot circulate. Pits or erected containers are totally inadequate.

In every situation it is important that the compost pile should not have direct contact with the soil (it must not be placed on gravel or paved places), thus small animals could enter through compost. Through this, ponding is also avoided. As a settlement place, the best place is protected against the wind and semi-shade. Therefore, the drying of the material from the pile is prevented. Not only do hedges protect well the compost layers but they are also, at the same time, a shield against the wind.

Building a compost pile starts with a layer of chopped branches with a role to protect against moisture. Then, the other materials follow in layers, well mixed, as much as possible. In general, the more the organic materials are diverse in composition, the more the compost quality will be better. What is important is that solid materials should be chopped, otherwise they hardly decompose. The rigorous mixing of materials helps aeration. In order to maintain the forming process of compost, an optimal degree of moisture is needed. A simple test is material for compost, pressed by hand, in such a way that water must not fall, but the hand must be wet.

The materials used for composting must be separated according to their content of nitrogen and how fast they go into rot. Tree branches shall be chopped and brought to the best dimension of 5–10 cm. Straws are mixed with fresh and humid manure.

The carbon/nitrogen ratio is very important and it is advisable that it should be of 20–40/1 (C:N) in the whole mixture.

The compost shall be rich in nitrogen if the following are used: liquid manure, bird manure, grass, vegetable scraps, cattle manure, kitchen waste, potato peel, horse manure, a layer of dead leaves of alder, ash, hornbeam, feathers and wool scraps.

The compost shall be rich in carbonates should the following materials be used: a layer of dead leaves of linden, oak, birch, peat, fruit, a layer of conifer needles, barley, oat, rye straws, tree bark, shrubs, bushes, sawdust, paper and carton.

A successful mixture will result if 3 parts of materials rich in carbonates and 1 part of materials rich in nitrogen are used.

The duration of the fermenting process depends on the care, the material and the size of the compost pile. Compost can be used in different stages of its development. An empirical rule tells us that, under optimal conditions, the fermenting process is finished after almost 6–9 months; in this case, the temperature is very important. Then the compost smells like fresh wood soil and presents itself as a dark brown, frail material, which can be sieved.

Compost should not be left for a long time without being used because if it gets old it loses its nutritional substances. The organic material decomposes the fastest during spring-autumn. During the cold months, the fermenting process stops.

Mainly, the compost resulted is useful for the entire garden. It must be set in a layer of five centimeters thick. Upon sowing carrots, lettuce, peas, radishes, etc., a thin layer is enough, while in the case of cabbage, celery, garlic, cucumbers, potato and other such vegetables, more compost can be used. Compost is indicated even for flowers and shrubs; in late autumn, when the grass is no longer mown, it can be applied on the grass in a thin layer. One single application of compost per year is usually enough.

Compost is not indicated as a direct material for plants that need a lot of moisture (azalea and rhododendron). Rusty or musty compost should not be used under any circumstances. If it smells bad, the compost has started to rotten. This indicates the oxygen deficiency or the storing of too many amounts of the same type (humid grass and leaves in thick layers). The pile must urgently be turned. If there are unwanted animals (rats or mice), it is because of leftovers, that is why leftovers must be immediately mixed with other materials and covered or a fine layer of calcareous stone should be spread above them. The appearance of molds in such a great number is usually caused by dryness. It is recommended that, upon turning the pile, the material should be wet.

A few rules to obtain a good compost:

- The material for compost shall never be put in a pit; the air cannot reach it; non-aired compost rots and smells bad.
- Hermetically sealed containers shall never be used as it results in an oxygen deficiency.
- The compost pile shall not be lifted on a stone, concrete foundation, etc.; worms have an important role, that is why compost must be set in permanent relation with the soil; still, if it cannot be set directly on the ground, worms must be added.

- Compost shall be set in the shade; for shade, average height shrubs, hedges or plants which grow directly on the compost pile are adequate (pumpkin).
- For the inferior layer, almost 20 cm thick, solid material shall be used; then, finer components, for example, leaves put in layers or mixed; grass (only dry) shall be spread in a thin layer; thus, there is a chance that putrescence should appear.
- On each layer one shall sprinkle a little bit of garden soil, compost or fertilizers based on a calcareous stone to stimulate ferment, then it is mixed and watered.
- The scraps which can attract animals are covered well with soil (meat, salami scraps, etc.).
- Drying the material from the pile shall be avoided; small animals need moisture, but compost should not be too watery either; in this case, oxygen lacks and putrescence follows.
- Onion peel, onion scraps, coffee and tea grounds are the ideal food for the animals from the soil.
- The compost pile shall be covered; thus, the heat release is stimulated and water or nitrogen losses are avoided.

When composting domestic waste, special attention must be paid to the following materials:

- Orange peel can compost, but the decomposing period is long and can last in time, that is why it is recommended that they should be used in small quantities.
- Red seeds shall not be used in compost, as they can sprout in the compost and germinate in unwanted places.
- It is better that the dry fruit peel should be burned and the ashes are used in compost.

The following should not be put in the bio-trash cans: fluid, cooked or salty letfovers, bones, cat or dog bed, hair, textile waste, glass, metal, batteries, chemicals of any kind, varnish, drugs, used oil, baby diapers, the content of the sack from the vacuum cleaner (danger of loading with heavy metals), straws, if the grains have been treated with pesticides or growing phyto regulators.

There is a great number of substances in stores to start and accelerate composting, prepared with bacteria and fungi, which 'light' the fermenting process in its early phase. These liquids are necessary only in certain situations.

COMPOSTING OF SAWDUST

The waste from the wood industrialization is valued more as basic material in the production of thermal energy, both in the private sector (small farms, households, the district heating system of blocks of flats, etc.) as well as for heating systems.

Heaps of sawdust and other wood waste are sources of aggressive pollution for forest soil and water streams. Therefore, the pollution with wood waste has the following consequences: removal from the productive circuit of certain areas of land on which vegetation disappears or reinstalls with difficulty, the normal circuit of surface water, the wind direction and the solar irradiance of the land modify. Also, climate changes take place at the same time with the development of bacteria, larvae, insects and fungi as well as the reduction of the vegetal carpet, the development of weed and the spread of wood powder into the atmosphere. Physical and legal persons who deal with wood processing shall store the waste resulted selectively on concrete platforms.

When obtaining compost from wood waste, a very wide range of organic matter must be used, and among the wood waste, sawdust is generally used.

For the composting of wood waste, adding also other organic waste is very important. The role of adding organic waste is to increase the share of nitrogen, phosphorus and hydrolysis intermediate products of hydrocarbon substances as well as proteins, vitamins needed for the development of the microflora which activates the fermenting process.

Table 6.7 The chemical composition of the main raw materials used for composting

Raw Materials	*Organic Substance,* %	C:N	N, %	P_2O_5, %	K_2O, %	CaO, %	MgO, %
			Communal Sector				
Domestic waste	20–80	12–20	0.6–2.2	0.3–1.5	0.4–1.8	0.5–4.8	0.5–2.1
Green waste	15–75	20–60	0.3–2	0.1–2.3	0.4–3.4	0.4–12	0.2–1.5
Waste water sludge	20–70	15	4.5	2.3	0.5	2.7	0.6
Paper – carton	75	170–800	0.2–1.5	0.2–0.6	0.02–0.1	0.5–1.5	0.1–04
			Solid Manure				
Cattle	20.3	20	0.6	0.4	0.7	0.6	0.2
Horse	25.4	25	0.7	0.3	0.8	0.4	0.2
Sheep	31.8	15–18	0.9	0.3	0.8	0.4	0.2
Pig	18–25	14–18	0.8	0.9	0.5	0.8	0.3
			Liquid Manure				
Cattle	10–16	8–13	3.2	1.7	3.9	1.8	0.6
Pig	10–20	5–7	5.7	3.9	3.3	3.7	1.2
Poultry	10–15	5–10	9.8	8.3	4.8	17.3	1.7
			Secondary Products				
Straws		50–100	0.4	2.3	2.1	0.4	0.2
Beet leaves 70	70	15	2.3	0.6	4.2	1.6	1.2
Fresh tree bark, sawdust	90–93	85–180	0.5–1.0	0.06	0.06	0.5–1	0.04–0.1
Mulch of tree bark	60–85	100–300	0.2–0.6	0.1–0.2	0.3–1.5	0.4–1.3	0.1–0.2
Grape marc	80.8	25–35	1.5–2.5	1.0–1.7	3.4–5.3	1.4–2.4	0.21
Fruit marc	90–95	35	1.1	0.62	1.57	1.1	0.2

The process of obtaining composts is, technically, well driven through the periodic inspection of temperature, moisture, pH value and content of total nitrogen, mineral nitrogen and organic matter. When executing the proper works of composting sawdust, one should take into account the physical-chemical features of the sawdust stored for a long time in free air and the one freshly obtained as the composting technology has certain differences.

In order to obtain a good compost, one shall choose carefully the materials which are mixed with the wood waste to correct certain deficiencies of sawdust in this regard. The chemical composition of the main raw materials used for composting is presented in Table 6.7.

From the table, it results that wood waste is among the weakest raw materials when it comes to the content of substances, which makes it hard, through separate ferment, to obtain a good compost. The content of nutritious substances (phosphorus and potassium) must be supplemented by their mixture in a sufficient quantity. As a consequence, by adding certain liquid or solid manure, the quality of the compost obtained can be significantly improved.

COMPOSTING PROCESS

Compost can be obtained through the ferment of different organic debris (straws, cocoon scraps, chaff, weed and vegetable scrap, degraded fodder, wood waste, domestic waste, etc.), to which we add sometimes mineral substances (lime, ashes, etc.). Gathered in piles, the debris must be managed from the point of view of aeration, temperature and moisture to favor the fermenting process (Pagans et al. 2006).

The composting process involves from a microbiological point of view a complex network of factors, which can be represented as a food pyramid with consumers at a primary, secondary and third level. The basis of the pyramid or the energy source is made up of the organic matter represented by plant and animal residues (Dumitrescu 2004).

Third consumers (organisms which feed with secondary consumers) are represented by myriapods, different species of bugs, spiders, ants, etc.

Secondary consumers (organisms that feed with primary consumers) are certain types of spiders, bugs, nematods, protozoa, worms from the soil, etc.

Primary consumers (organisms which feed with organic scraps) are bacteria, fungi, actinomycetes, nematods, certain types of bugs, worms, myriapods, snails, etc.

Organic residues are represented by leaves, grass and other plants' scraps, food scraps, scrap feces and animal scraps.

By analyzing the food pyramid, one can notice that organic scraps are consumed by certain kinds of invertebrates (myriapods, snails, shellfish, etc.). These invertebrates chop the organic matter creating thus a bigger, available surface against the attack produced by fungi, bacteria, actinomycetes which, in their turn, shall be consumed by organisms, such as spiders, certain types of bugs and worms.

Many species of worms, including earthworms, red worms, nematodes feed with decomposed vegetal debris and with germs, excreting in exchange organic compounds which enrich the content of compost, creating an aeration effect (due to the tunnels practiced) and increase the contact surface of organic matter for the subsequent action of germs.

With every consumer who dies or produces organic substances following the excretion processes, the number of food increases for the rest of the organisms which produce the decomposing of the organic matter.

Composting is generally defined as a decomposing process by biological oxidation of the organic constituents from waste, practically of any kind, under controlled conditions. As composting is a biological process of decomposing of the organic matter, it needs special conditions, mainly determined by optimal values of temperature, moisture, aeration, pH and C:N ratio, needed to ensure the best biological activity in the different stages of the composting process (Dumitrescu 2004). The main products of the composting process are carbon dioxide, water, different mineral ions and stabilized organic matter called humus or compost.

THEORIES ON THE CONSTITUTION OF THE HUMIFIED ORGANIC SUBSTANCE

The humification of organic debris is the sum of processes which take place under the action of enzyme organisms and complexes through which subproducts result out of very diverse organic compounds, which by condensation form 'humus' (Blaga et al. 2005, Szajdak et al. 2017).

The organic substance of the soil is made up of material preserve and potential energy resulted from the biological processes which take place in it, as well as of materials and secondary products of these processes. The animals from the soil form 'the vital part', while the humus forms the 'postmortal' part. The organic substance of the soil can be divided into non-humified materials and humified materials. The materials without humus are formed from vegetal organisms and destroyed animals and products from their decomposing in the process of humification. The humified matter is represented by stable organic compounds with big molecular dimensions. Humus is an organic colloid of soil. Its feature lies in a big specific surface and the ability to reversibly fix molecules and water ions, defining the adsorption ability of the soil. It has an important role in forming the soil structure, in the adsorption of nutritious substances and the soil water and heat regime.

To a greater or lesser extent, humus can be dissolved from the soil with basic solvents (NaOH, N_2CO_3), liquids of neutral salts (NaF, $Na_4P_2O_7$ and salts of organic acids) or chelating agents. Based on their behavior toward the diluted base and liquid acids, we can classify the humic materials into three main groups:

- Fulvic acids, which dissolve in acid and bases.
- Huminic acids, which do not dissolve in acid, only in base.
- Humines, insoluble in bases and acid.

The main features of the humic materials are presented in Table 6.8.

Table 6.8 The main features of the humic materials

Specific Designation	*Fulvic Acids*	*Huminic Acids*	*Humines*
Molecular volume	2,000	5,000–100,000	300,000
C,%	40–50	55–60	55–60
N,%	<4	4	>4
O,%	45–48	33–36	32–34
Structure capacity		→ increasing →	
Solubility in water		← decreasing ←	

Humus does not have a unitary structure, it can be characterized only as a complex group of compounds. The fact that humic materials are amorphous compounds, so they cannot be crystallized, makes it hard to know in detail their structure. Humus has a polymeric structure, set up of basic, monomeric elements. The stability of the humus compounds increases proportionally with the number of skeleton compounds from the core.

The functional groups can be acid (ex. groups of de carboxyl, hydroxyl) and alkaline (ex. groups of methoxyl, amino-nitrogen). The capacity to react to these groups ensures the chemical formation of humus materials. Around functional groups hydrate membranes form. It often happens that polymerized non-humified materials through functional groups should attach themselves to humified materials, and what is characteristic for them is their easy untying from the humus materials and their decomposing by the microbial route. But basic aromatic elements resist the microbial decomposing and from a chemical point of view, they are very stable compounds.

We can divide decomposing (fragmentation and mineralization) and the set up of the humified organic substance into three stages:

Introductory biochemical phase (initially). In this phase, in the hydrolytic and oxidative processes, polymers with big molecules decompose into dimers and monomers. This decomposing in the tissue structure is not easily visible, a good example is the browning of leaves.

Mechanical fragmentation. Following the edaphic effect, organic substances fragment and mix with mineral components.

Microbiological decomposing. Organic compounds decompose by enzymatic route into constructive elements. The heterotrophic and saprophyte organisms are responsible for this process. The oxidation of organic substances—mineralization—releases mineral substances, water, carbon dioxide and energy.

Lignin has a very important role in the syntheses of humus materials, as it is formed mainly by the condensation and polymerization of lignin decomposing products.

When it comes to the humus formation there are several theories:

1. The lignin theory: Humus materials are formed from modified lignins in such a way that they lose the methoxy groups and hydroxyl groups are formed by oxidation on aromatic skeletons, while on the side chains carboxyl groups are formed.
2. The polyphenol theory: Products that are formed during the microbial decomposing of vegetal cells and secondary products of the microbial metabolism are transformed through polymerization, resulting in humus materials. The lignin from the cell wall decomposes by the oxidative route and polyphenyl–propanol monomers are released. During the subsequent decomposing, the side chains break and the aromatic rings lose their methoxy groups. The polyphenols formed in this process oxidate into quinones. By reacting later, for example with amino acids, quinones form condensation products. The initial compounds of humification are polyphenols, which can come from the metabolism of germ cells and the cell autolysis.

3. The melanoid theory: melanoids are formed from saccharines and amino acids. This reaction is called the 'Maillard reaction'. Following the common transformation of decomposing products of polysaccharines and proteins, the melanoid product is born.

The humus materials are formed under the influence of microorganisms (biochemical synthesis). This process can also run in several ways:

- In the cell of certain microorganisms, humus materials are formed.
- By transforming different microbial substances during the cell autolysis.
- With the help of ecto enzymes, live microorganisms oxidate aromatic compounds from the sublayer.
- Forming humus materials from the products resulted from cell autolysis.

In the biological humification of organic substances, microorganisms have an important role. The explanation for this lies in the high density of the microorganism population and in the high intensity of their metabolism.

The formation of humic materials can take place under biotic and abiotic conditions. The biological humification takes place relatively fast under the conditions of a neutral pH and following a strong microbial effect (He et al. 2013). In the abiotic humification, microorganisms participate only at the beginning of the process. Later, the process moves on slowly, accompanied by an acid chemical reaction.

Humification is a complex process and many of its elements are not known fully not even today. Based on the dominant processes in the formation of humus materials, we can distinguish among four phases.

- *The metabolic phase*: The biogenesis of aromatic compounds; partial decomposing of aromatic compounds.
- *The phase of radical formation*: The formation of 'extreme' compounds (formation of humus forerunners of the humic acid).
- *The conformation phase*: Humus forerunners of the humic acid tie with the non-aromatic compounds.
- *The phase of formation of the humus system*: Tying humus materials with mineral parts, forming organo-mineral complexes.

The humified materials can be divided into forerunners of the humic acids (fulvic acids) and humines, which can be considered as final products of humification.

During the first phase, microorganisms are active. During the radical phase, microorganisms have a more reduced role and during the last two phases, they do not have any influence anymore (based on current results). The processes run successively, but for a shorter or longer period, the process can also stagnate.

From the point of view of the soil biology, composting is a process which can be identified with rot, during which, with the help of aerobic microorganisms, organic substances mineralize, respectively, in a certain proportion they humify, having compost as a final product that is nothing else but the sum of stabilized (humified) organic substances of mineral nutritious substances and microbial products (ferments).

During composting, organic substances suffer the following quality changes:

- The C:N ratio decreases.

- During composting, the content of organic substances with light solubility is reduced (carbon hydrates and proteins).
- The quantity of humus forerunners is decreased (fulvic acids) and the quantity of huminic acids is increased.
- Changes can be noticed in the quality of organic substances (the number and quality of functional groups), which are similar to the humification in the soil.

Quantity and quality changes of the organic substances which appear during composting have the following meanings:

- A stable organic substance forms in the well-conducted process, which resists to microbial decomposing; due to the stability of the organic substance, storing compost has small risks for public health compared to raw organic substances.
- Reaching the soil, they do not induce unfavorable processes to the soil biology (rot).
- Through the effect of the quality changes occurred (color, conditions of adsorption and level of polymerization) they favor the physical and chemical features of the soil.

The decomposing by microorganisms and the incorporation of mineralized substances in the construction of new humified materials during the composting run in parallel processes. The energy needed for all this system is obtained through the oxidation of organic substances. In the presence of $O_{2,}$ the materials which supply energy with carbon content oxidate in CO_2. In compost, aerobic relations must also prevail for more efficient energy use, although anaerobic transitory areas can always form, where oxygen-free relations can be observed and among others, the formation of CH_4. By choosing the composting materials, we must ensure the best oxygen quantity as if there is too little oxygen, anaerobic relations install and excessive supply with oxygen lead to the cooling of the prism. The requirement of O_2 changes once the material transforms and reaches its highest when the intensity of the activity of microorganisms is the highest, respectively, when we ensure nutritious substances, humidity and adequate temperature for its functioning.

Nitrogen behavior in composting: Nitrogen is a constitutive element of amino acids, of proteins and of nucleic acids as well as of chlorophyll and being at the same time the important constitutive part of the bearer of life, of protoplasm, of chromosomes and cell elements and genes, which store and transmit genetic information of ribosomes too.

Nitrogen is made up of a series of organic and inorganic compounds, which are very important for the chemical reactions which take place in compost as a biological system. In nitrogen chemistry, the most significant is the redox processes, associated with the transport of electrons. The content of nitrogen of compostable waste is to a great extent fluctuant. The composition of communal waste changes a lot according to season and the collecting system.

The first important process during composting is the mineralization of nitrogen compounds, whereas first step nitrogen as ammonium is released by heterotroph microorganisms through the ammonification process. Among microorganisms that make the ammonification, there are species that degrade only their proteins, nucleic acids, etc., but there are also species with a wide range of mineralization.

The release of ammonium from amino acids takes place through deamination. It has two variants:

1. The combination of the transamination with the oxidative deamination of glutamate.
2. Direct oxidative deamination.

In the first case, the amine groups transform into ketoglutarate. In the second case, the amino acid is deaminated unspecifically by flavinenzyme. H_2O_2 formed simultaneously is decomposed by catalase.

MICROORGANISMS INVOLVED IN COMPOSTING

In the composting process, microorganisms make the splitting of the organic matter producing carbon dioxide, water, heat, moisture and humus as a relatively stable final product (Ravindran et al. 2019).

Under the best conditions, composting is achieved in three phases (Dumitrescu 2004):

a) The mesophilic phase or of moderate temperature, which takes a few days.
b) The thermophilic phase or of high temperature, which can take from a few days to a few months.
c) The cooling and maturation phase, which can take several months.

During the three composting phases, different communities of microorganisms prevail. The initial decomposing is ensured by the mesophilic microorganisms, which rapidly decompose the soluble organic compounds, easily biodegradable (Vargas-Garcia et al. 2010).

The biochemical reactions of decomposing are exothermic, therefore the temperature of compost rises. When the temperature exceeds 40°C, mesophilic organisms become less competitive and are replaced by the thermophilic ones and are resistant to heat. At temperatures of 55°C or higher, many microorganisms, among which human or vegetal pathogens are destroyed. As temperatures over 65°C cause the death of many forms of microorganisms and limit the decomposing speed, one can proceed with the aeration and mixing of compost to maintain temperature under this value.

During the thermophilic phase, high temperatures accelerate the splitting of proteins, fat and polysaccharides, cellulose and hemicellulose and the structural organic compounds mostly found in plants. Once with the biodegrading of these rich organic compounds which take in their structure, a high quantity of energy, the temperature of compost decreases gradually and the mesophilic microorganisms intervene again, producing the final phase of maturation of the organic matter remained undegraded.

The formation of compost is an exothermic process, the energy which produces becomes free as heat. Based on the change of temperature in the process of formation, we can distinguish among four phases:

- The introductory phase.
- The decomposing phase.
- The transformation phase.
- The recovery phase.

Under the best conditions of the first short introductory phase, microorganisms begin to multiply at high speed. With the effect of intensive metabolism, the temperature rises rapidly to the thermophilic area. The duration of the introductory phase is generally a few hours, eventually 1–2 days. One must mention that the importance of the introductory phase practically and theoretically is to be neglected as a consequence most of the authors do not even mention it as a phase as such.

The mesophilic microorganisms are responsible for the decomposing or thermophilic phase for which the best temperatures are 25–30°C and due to their intensive metabolism, the temperature continues to rise. The number of mesophilic organisms increases up to 45°C, over 50°C they already destroy in a great number and over 55°C only their forms resistant to temperature are maintained. All this needs 12–24 hours. Once with the destruction of the mesophilic microflora, the thermophilic microorganisms multiply fast for which the best temperatures are 50–55°C. But certain species remain active even at 75°C. Over 75°C there are no more biological processes, but what is more characteristic are the purely chemical autooxidative and pyrolytic processes.

Between the mesophilic and thermophilic microorganisms, there is metabiosis. The heat produced following the metabolism of mesophiles ensures the temperature needed for the requirements of the thermophilic flora and through their transformation activity of organic substances, they ensure the thermophilic microorganisms better access to the nutritious substances.

The decomposing phase can last even several weeks. After this phase of ferment, the temperature drops significantly.

In the transformation phase, microorganisms begin the decomposing of lignins hard to decompose, and mono-, di- and triphenol compounds are formed during this process. Humus molecules result from their condensation.

The last phase of composting is the recovery phase, which is characterized by the humification (caramelization) of organic substances, gives a dark color to its final product. Then one can notice the decrease in temperature of compost. Bacteria, for which the best temperatures are 15–20°C, mainly take part in ferment. In this phase, the number of actinomycetes grows significantly, which can also be the indicator for the maturation of compost.

During composting, according to the raw materials used, the environmental conditions and the degree of ferment, different animals take part in the decomposing, transformation processes and processes to recover organic materials. The compost shall be populated by microfauna only towards the end of the ferment. First, its chopping, selection and mixing activity determines the physical characteristics of the mature compost. From the point of view of the compost maturity, microorganisms have an important role. Microorganisms taking part in ferment can be divided into the following groups:

- Aerobic and optionally anaerobic bacteria.

- Actinomycetes.
- Fungi.
- Algae and protozoa (single-cell).

Bacteria are the smallest and the most numerous live organisms which activate in compost, representing 80–90% of the millions of typical microorganisms of a gram of compost. They are single-cell animals, 10–30 μm in diameter. Their most remarkable feature is that they do not have a cell core delimited by the membrane. The DNA is found in the plasma, freely, in a single chromosome. They do not have mitochondria and chloroplasts. They have less varied forms, we can distinguish between two forms: comma (vibrio) and rod (bacilli). Their density is 1.07 g/cm^3 and because of their small dimensions, the surface specific to the cell is very big, which makes it possible for a highly intense metabolic activity. Under adequate conditions, certain strains can decompose in one hour a quantity of glucose 100,000 higher than their mass. The bacterion cell contains 80% water and 20% dry substance. By cell division they are capable of a fast reproduction, a generation's time—in the best case—is no longer than 20 minutes. From the point of view of their metabolism, one can distinguish between autotrophic bacteria—which use carbon dioxide from the air, similar to the higher-order vegetal organisms—and heterotrophic bacteria—which use organic compounds as a carbon source. During composting, the chemoorganotroph organisms are the most important, where organic compounds have the role of a hydrogen donor.

Bacteria are responsible for most of the part of the decomposing process and for generating heat in compost. They represent the most diverse nutritional group, which activates in compost using a large field of enzymes to split a big variety of organic materials.

Bacteria are single-cell organisms structured into sphere bacilli, sphere or spiral cocci, capable of their movement. Mesophilic bacteria prevail at the beginning of the composting process (0–40°C); many of them can also be found in the soil. When the temperature of compost exceeds 40°C, thermophilic bacteria, the population dominated by the *Bacillus* type replace the mesophilic ones,. The diversity of the bacilli species is big at temperatures between 50–55°C, but it dramatically decreases at temperatures over 60°C. When the temperature becomes unfavorable, bacilli survive by forming endospores, spores with thick cell walls which are very resistant to heat, cold or lack of food. They stay in nature and become active when environmental conditions become favorable.

At high temperatures, bacteria from the *Themus* family were isolated from compost. Once the cooling process of compost begins, mesophilic bacteria start prevailing again. The number and type of mesophilic bacteria which recolonize compost upon maturation depend on the spores and microorganisms present in compost as well as on the environmental conditions. In general, the longer the maturation process, the more diverse the microbial community.

Actinomycetes: Their fitting into the system is part of the controversial problems of microbiology. Some consider them rather bacteria and others between bacteria and fungi as an independent group. Their enzyme systems can decompose lignin, they have an important role in the humus formation and its mineralization. During their metabolic process, they produce antibiotics and vitamins through which they have a significant role in the biochemical cleaning of mature compost and forming the stimulating effect of plant growth. The smell similar to forest soil of ferment compost is due to the actinomycetes which live in it. Like bacteria, actinomycetes do not have a core, but they grow like multicell filaments similar to fungi. They play an important part in the composting process in biodegrading complex organic substances like cellulose, hemicellulose, lignin, chitin or proteins.

The enzymes they secrete make possible the biodegrading of wood debris, tree bark or newspaper. Certain species appear during the thermophilic phase, others become important during the maturation phase when they remain to form humus.

Actinomycetes form long, branched filaments, like a spider web in compost. One can notice these filaments towards the end of the composting process, sometimes as a circular colony which gradually increases in diameter.

Fungi belong to eukaryotes and are similar to higher-order plants. Under aerobic conditions, fungi meet the energy demands by oxidating organic substances. They can decompose the wood vegetal parts with a high content of cellulose and lignin, they can synthesize reserve nutritious substances, for example,: fat, polysaccharides, organic acids, vitamins and antibiotics. Fungi which take part in decomposing cellulose during composting, at temperatures that exceed 60°C are especially important. They become visible during composting when white micelles intertwine the external dry surfaces of compost.

Fungi are responsible for decomposing in the soil and compost of several complex polymers from the plant structure. In compost, fungi are important because they split cellulose waste allowing thus for bacteria to continue the decomposing process. They develop by producing many cells and filaments and can attack organic residues that are too dry, acid or with low content of nitrogen in order to be decomposed by bacteria.

Many fungi can be classified as saprophytes as they live colonizing dead organisms, obtaining energy by splitting organic matter from plants or dead animals.

The fungi species are numerous both in the mesophilic phase and in the thermophilic phase of composting. In general, they can be noticed in a great number during the storing of the fermented compost.

Protozoa are single-cell microscopic animals, which live in the water drops from compost. Their role in the composting process is relatively minor. They take their food in the same way as bacteria do, but they act as secondary consumers by feeding with bacteria and fungi.

Under the influence of microorganisms, the composting process runs as follows:

1. Microorganisms present in organic waste and/or in the air begin to decompose materials, a process accompanied by heat release, thus the temperature of the pile rises. When organic acids (lactic acid, butyric acid) are produced, the pH level starts to decrease.
2. Over 40°C, the decomposing process begins. The temperature rises to 60°C; here fungi are not active, only the species of actinomycetes and bacteria are. At this high temperature, compounds by decomposing (sugar, starch, fat and proteins) are the ones that consume faster. By releasing proteins

from ammonia, pH becomes alkaline. In this phase, the decomposing of the more resistant material (cellulose) begins, the speed of reaction is reduced, being accompanied by a drop in temperature.

3. Once the temperature falls, thermophilic fungi from the pile multiply again and contribute to the decomposing of cellulose. Later, microorganisms from phase 1 reactivate again, too. This process develops relatively fast, during a few weeks.
4. The ferment is the last phase and it needs several months. The reactions which take place lead to the transformation of the organic matter remained in big humus molecules or huminic acids. In this phase, there is a big competition among microorganisms for nutritious substances. Representatives of microfauna (mites, ants and worms), which take part in the chopping of organic debris make their appearance in the pile.

INFLUENCE OF THE SUBLAYER PROPERTIES UPON FERMENTATION

In order to obtain a good compost, one must be aware of the features of the organic materials used and how they influence the fermenting process. These features are (Goyal et al. 2005, Solti et al. 2006, Villar et al. 2016, Foereid 2019):

1. The chemical composition (content of organic substances, C:N ratio and content of nutritious substances). In order to be composted, the materials used must have at least 30% organic substances. Another important characteristic is, as was mentioned before, the C:N ratio, the optimal value of which is 25–30:1 and which can usually be obtained by mixing raw materials. The content of other nutritious substances (phosphorus and potassium) is less important for the composting process because they, in general, are at the disposal in the quantity needed for ferment or in case they lack, they can easily be supplemented before composting. Some materials used can be very rich in a certain nutritious substance, so their application can lead to the improvement of the compost quality. For example, the marc of grapes, which forms in big quantities in the winery is rich in potassium and phosphorus.

Carbon and nitrogen are the most important of the multiple elements of the composting process. Carbon ensures both the energy source as well as 50% of the structural material of microbial cells. Nitrogen is the main component of proteins, nucleic acids, amino acids, enzymes and coenzymes needed for the growth and functioning of microorganism cells. With low values of the C:N ratio, nitrogen in excess shall be transformed in ammonia gas, which gives compost an unpleasant smell. Bigger values result in insufficient quantities of nitrogen for the best growth and development of the microbial population, therefore compost shall remain relatively cold and biodegrading will occur at low speed.

The humid and green vegetal materials are rich in nitrogen, while the dry, brown materials are richer in carbon. The C:N ratio of the composting mix can be calculated and adjusted by using a combination of materials very rich in carbon and nitrogen (Dumitrescu 2004):

Materials rich in carbon (C:N ratio):

- Dry leaves (30–80:1)
- Straws (40–100:1)
- Wood debris, sawdust (100–500:1)
- Tree bark (100–130:1)
- Debris of mixed paper (150–200:1)
- Newspaper and carton (560:1)

Materials rich in nitrogen (C:N ratio):

- Vegetal debris (15–20:1)
- Coffee debris (20:1)
- Grass and green plant debris (15–25:1)
- Manure (5–25:1).

Once the composting process advances, the C:N ratio gradually decreases from 30:1 to 10–15:1 in the final product. This is due to the fact that 2/3 of the carbon consumed by microorganisms is transformed into carbon dioxide and then eliminated. The third part of the quantity of carbon left is incorporated together with nitrogen into microbial cells and made available again for reuse after the cells die.

As the organic compounds from the waste components resist microbial decomposing in different ways in order to obtain the best composting dynamics when mixing raw materials it is not enough to take into account only the C:N ratio. For example, carbon is slowly released from raw materials rich in lignin (e.g. sawdust) and if one adds nitrogen sources with rapid decomposing to these, serious nitrogen losses appear as ammonia, which is also a pollution source besides an economic loss.

2. The compostability of materials used: As organic compounds from waste have a different resistance towards microbial decomposing in order to obtain the best dynamics of degrading when materials are mixed, one must also take into consideration the different suitability to decomposing. If the materials used are very rich in lignin (for example, sawdust), carbon is slowly released and if we add nitrogen with rapid decomposing to these materials, then a serious nitrogen loss appears as ammonia, which is also a pollution source besides an economic loss.

Although maintaining a C:N ratio of 30:1 is one of the conditions of the composting process, this ratio can be adjusted according to the availability of composting materials.

Besides the C:N ratio, there are also other chemical elements like phosphorus, potassium, as well as tracks of minerals (calcium, iron, boron, copper, etc.), which are essential for composting and the metabolism of microorganisms. Normally, these elements are not limited as they are present in relatively high concentrations in the composting materials.

3. Structural stability, particle size and loosening are other features of the materials used which influences composting. Out of a basic material with a poor porous structure during the ferment, oxygen consumes quickly, making way for unfavorable anaerobic processes. During composting, the minimum porosity (loosening) is 30% of the volume, which we can ensure by mixing structural

elements in an adequate quantity. Such basic materials with good structures are green residues, straws and wood chips (Table 6.9).

Table 6.9 The most important features of raw materials

Raw Material	***Structure***	***Humidity***	***Mixing Ratio***	***Requirement of Preliminary Treatment***
Biowaste	good-bad according to origin	high-average according to origin	50–100%	chopping, homogenization
Green waste	good	dry-average	0–100%	chopping, homogenization
Residual water sludge	bad	very high	max. 30%	–
Carton paper	good	dry	max. 60%	chopping
Stable manure	bad-average	average	max. 80%	mixing
Liquid manure	bad	very high	20–60%	mixing, dehydration
Straws	good	dry	max. 50%	chopping
Tree bark, sawdust	good	dry	0–100%	chopping
Beet leaves	bad	average	max. 50%	–
Marc of grapes	bad	high	30–60%	–

Oxygen is essential for the metabolism and breathing of microorganisms as well as for the oxidation process of different organic molecules present in the composting material. At the beginning of the microbial oxidative activity, the oxygen concentration in compost is approximately 15–20% (similar to the normal composition of air), and the carbon dioxide concentration is 0.5–5%. As the biological process progresses, the number of oxygen drops and the amount of carbon dioxide increases. If the oxygen concentration falls under 5%, the decomposing process becomes anaerobic. Without a sufficient amount of oxygen, the process becomes anaerobic, hydrogen sulfide is produced, which gives compost an unpleasant smell. By maintaining the anaerobic activity at a minimum, the composting reactor shall behave like a biofilter, degrading the bad-smelling products formed as secondary products of the anaerobic processes. The maintenance of aerobic conditions in composting can be done by using different methods, like practicing ventilation holes, introducing ventilation ducts, forced air currents, agitating the composting mixture and increasing its porosity.

Oxygen concentrations higher than 10% are considered the best in order to maintain aerobic composting conditions.

The decomposing of the organic matter first appears at the surface of particles, where the oxygen diffusion into the water pellicle which covers the particles is adequate for the aerobic metabolism and the sublayer is accessible to microorganisms and their enzymes. Small particles have a surface area per mass or volume unit higher than the bigger ones, and they will biodegrade faster should aeration be good. Particle sizes of 1,3–7,6 cm are recommended.

4. Compost humidity: The content of humidity of the materials used can be different. Neither too dry raw materials nor too humid ones are recommended.

The humidity needed (40–60% of the volume) to start the composting process is ensured by the mixture of materials and moistening.

The decomposing induced by microorganisms runs very fast in the fine liquid layer from the surface of the organic particles. An amount of humidity less than 30% inhibits the bacterial activity, while one higher than 65% shall result in a slow decomposing, production of unpleasant smells due to the formation of anaerobic decomposing areas and washout of nutrients from the mixture for composting.

The content in the humidity of compostable materials differs a lot based on their chemical structure. For example, it is 80% in the case of peaches, 87% in the case of lettuce, but 10% in the case of dry food for dogs and 5% in the case of a newspaper. Often, materials very rich in nitrogen are very rich in humidity, while the ones rich in carbon are dry. By combining different types of materials, a mixture that can compost efficiently can result.

5. Temperature: The speed of the composting process depends both on chemical factors and on physical ones. Among these factors, the key parameter to ensure the best composting is temperature. The heat resulted during the composting process is a secondary product of splitting organic matter by microorganisms. But it is one of the parameters which indicates the stage of the composting process. The heat production depends on the size of the composting installation, the content in humidity, aeration and the C:N ratio as well as the temperature of the environment.

Decomposing runs the most rapidly during the thermophilic phase of composting (40–60°C), which lasts from a few weeks to a few months, according to the size of the system and the composition of ingredients. This stage is also important for the destruction of thermosensitive pathogenic bacteria, insect larvae and weed seeds. In open systems, invertebrates from compost can survive during the thermophilic phase, for example, by migrating to the outskirts of the installation. Most of the species of microorganisms cannot survive at temperatures higher than 60–65°C, which imposes the agitation or aeration of the mixture to be composted if the temperature rises above this limit.

When the composting mixture starts cooling, the temperature can be increased again by overturning the composting reactor, which will favor the introduction into the system of a new amount of oxygen to provoke the decomposing of the organic matter left non-biodegraded. Upon the end of the thermophilic phase, the temperature of compost drops and cannot be increased anymore by agitating the mixture or overturning the composting reactor.

At this moment, decomposing is taken over by the mesophilic microorganisms and runs as a long maturation process of compost. Although during this phase the temperature of compost is close to that of the environment, chemical reactions continue to unfold, which results in the growth of stability of the organic matter left and the improvement of its nutritious qualities for the use in agriculture.

Temperature, anytime during composting, depends on the microbiological activity, the losses by conduction, convection and radiation. The smaller the composting reactor, the higher the surface-volume ratio, which will trigger a bigger loss of heat by conduction and radiation. The effect cannot be counteracted by isolating the reactor. The content of humidity also influences the compost temperature as water has a specific heat higher than other materials. Dry compost will tend to warm and cool faster than the one rich in humidity, ensuring an adequate level of humidity needed for the development of microorganisms.

6. The pH of material: The activity of the microorganisms involved in the composting process is the best when the values of pH range between 5.5–8.5. In the biodegrading process of the organic matter by bacteria and fungi, organic acids are formed, which during the primary stages of composting can accumulate, leading to the decrease of the pH value. The decrease of the pH results in the development of fungi, which shall biodegrade lignin and cellulose. Normally, organic acids are split in their turn during composting. But if the system becomes anaerobic, organic acids can accumulate and they can lead to the decrease of pH up to the value of 5.5, limiting severely the microbial activity. In such cases, it is enough to air the system to maintain the pH value under acceptable limits.

7. Preparing materials for composting: Before composting, a part of the raw materials needs certain treatments, the most frequent are grinding, chopping, pressing, homogenization, eventually removal of foreign materials. The reduction of the germinative capacity of weed seeds which are found in the materials used for composting is done by turning compost several times and running the material entirely through the thermophilic phase.

8. Auxiliary materials used during composting: Upon mixing raw materials, different additives can be added, which can influence the fermenting process and can improve the quality of compost. These can be ground clay, rock flour, lime, bone, blood and horn flour, etc.

9. Mass and volume of raw materials: It determines the composting process. By knowing these two properties, we can dimension the transport capacity and the surface needed for composting. The maximum mass per volume unit during ferment is 700 kg/m^3. The mass of the volume of raw materials is closely related to humidity, respectively their structure and the forming conditions.

Compost is a perfect source of nutritious substances for the agricultural and horticulture production; it is a material for soil improvement, protecting at the same time the plants against drought and disease. Preparing compost is simple, and its use increases the quantity and quality of production.

PRESENCE IN COMPOST OF POLLUTANT MATERIALS

During composting, the presence of pollutant materials must not be neglected. Based on the chemical properties, we can distinguish between organic and inorganic pollutant substances.

Inorganic substances are toxic heavy metals such as cadmium (Cd), chromium (Cr), copper (Cu), mercury (Hg), nickel (Ni), lead (Pb) and zinc (Zn). Their quantity is relatively small, but they can be toxic even in a small quantity, during composting they do not decompose and when they return to the food chain soil, plant, animal and human, they accumulate and reach the products from human consumption in concentrations harmful for our health. In the environment, these can be found anywhere (Table 6.10, Amlinger 1993), but their harmful effects can be avoided by meeting the approved minimum limits.

Organic pollutant materials are the chemical substances used currently (especially phytopharmaceutical substances), some of which are toxic, while others are very persistent. What makes them even more dangerous is the fact that in many cases their disintegration products are even more toxic and their decomposing process has not been known so far. The most dangerous organic compounds are polychlorinated biphenyls (PCB), polyaromatic hydrocarbons (PAH), polychlorinated dibenzo-p-dioxin (PCDD), polychlorinated dibenzofurans (PCDF) and chlorinated pesticides. During composting, special attention is paid to the origin of raw materials and thus the accumulation of pollutant materials can be avoided.

Table 6.10 Content of heavy metal of raw materials mg/kg in dry substance

Material	*Cd*	*Cr*	*Cu*	*Hg*	*Ni*	*Pb*	*Zn*
Soil	0.1–0.3	20–40	20–30	0.05–0.1	20–40	20–30	60–80
Plant	<0.1–1	<0.1–1	3–15	<0.1–0.5	0.1–5	1–5	15–150
Green waste	0.05–0.63	2.8–11.3	5.0–23.5	0.03–0.54	2.3–8.3	5.6–67.5	29.3–390
Stable manure	0.1–0.6	3–20	20–40	0.1–0.1	5–50	5–10	100–300
Vegetable scraps	0.2–0.4	1.5–2.5	3.0–7.0	0.01–0.02	2.0–4.0	1.0–2.0	70–100
Tree bark, sawdust	0.1–2.0	500–1ooo	10–30	0.1–0.3	30–60	50–100	40–500
Waste water sludge	3–17.8	64–572	190–863	1.8–6	37–202	145–600	1,320–3,232

Foreign substances for compost are the materials which do not decompose during composting, they are not toxic, but they decrease the value of the final product.

AUXILIARY MATERIALS USED DURING COMPOSTING

Upon mixing raw materials, different additive materials can be added, which can influence the fermenting process and can improve the quality of compost.

The additive materials can be ground clay; rock flour; lime; horn, bone and blood flour; urea and rice straw (Yang et al. 2019), liquids that favor composting.

Ground clay can be bentonite, alginite, illite, kaolinite. Clay mineral can have several functions in compost:

- Thanks to its high adsorption capacity, it diminishes the losses of nutritious substances and it has a very high capacity to fix ammonia.
- Due to its swelling capacity, it can reduce high humidity.
- With the humus compounds which are formed during the ferment, they compile clay-humic complexes, indispensable for the formation of stable soil glomeruli.

Usually, the dose of ground clay is approximately 20 kg for one m^3 of raw material. Of all ground clays, the most famous is bentonite, which can be found in stores.

Rock flour can be basalt, calcareous stone, dolomite, zeolite, rhyolite, etc. Materials of mineral origin have several positive effects upon composting:

- They increase the content of mineral substances of the finished product.
- They fight against germs.
- They reduce nitrogen losses.

What is characteristic of rock flour is its richness in different mineral substances. During composting, under microbial effect, these substances reach the compost by absorption, which is extremely important for the supplementing of composts with microelements.

One can also notice that rock flour has beneficial effects upon the ferment, which explains by several factors:

- Thanks to its effect of improving structure, rock flour ensures better oxygen supply.
- They have a capacity—similar to clay—of fixing excessive humidity.
- The microelements are released to stimulate microbial activity.
- They have the capacity to buffer and improve pH.

The usual dose of rock flour during composting is 20–40 kg/m^3, but it changes according to the quality of raw materials and the purpose of use.

Lime: In the past, all forms of lime were ingredients indispensable for composting, which in most cases had more negative effects upon the quality of compost. The adjustment of the chemical reaction does not need lime in most of the composts. Exceptions are only branches, plant leaves rich in tannin, wood waste, sawdust, their splinters and certain peat. The other raw materials can also have neutral or weak alkaline chemical reactions without lime.

Applying lime makes more sense when raw materials are poor in calcium. The lime added as a supplement cannot be by any means unslaked lime because the latter is accompanied by heat release reacts with water, and the sudden temperature and pH growth result in huge losses of ammonia.

For acid raw materials, it is recommended to supplement compost with calcium carbonate ($CaCO_3$), limestone or carbonate marl.

Horn, bone and blood flour: These additive materials rich in nutritious substances—bone flour contains a lot of phosphorus, and blood and horn flour contain a lot of nitrogen—can be used if the raw materials lack macro and microelements. But these materials are pretty rare and expensive.

Liquids that favor composting: Liquids that accelerate the ferment and blur mold are part of this group. Their common feature is that they are usually used in a very small amount.

There are two main groups of liquids:

- Herbal products
- Inoculant materials

One must underline that a good compost can also be obtained without these liquids. But this does not exclude the fact that certain substances have positive effects upon the ferment.

Herbal products: These products are especially grounds from medicinal herbs or their extracts. The most famous are products used in biodynamic agriculture, and they are produced by a very special method of the following plants like milfoil (*Achillea millefolium*), nettle (*Urtica dioica*), chamomile (*Matricaria chamomilla*), dandelion (*Taraxacum officinale*), allheal (*Valeriana officinalis*) and oak bark (*Quercus robur*).

As they have homeopathic effects, these materials must be added in the composting process in minimum quantities (1–2 g/m^3). According to biodynamic farmers, applying these liquids has the following effects:

- A smaller loss of nitrogen.
- Preventing putrefaction.
- Refined decomposing.
- A better success of cosmic effects during the ferment.
- A better quality of compost.

Inoculated products, starter crops: By these designations, we understand products rich in germs of different bacteria and fungi applied in order to accelerate the ferment of compost and favor the humus formation. The microorganisms which take part in the composting process can be found anywhere in nature. If in the composting process adequate conditions are created, microorganisms shall multiply fast and the best decomposing shall take place as long as the conditions favorable to the ferment are met.

By meeting all the indications needed to apply inoculant products, still a better composting is obtained. Inoculant products are needed especially in the case of special waste, very rich in oils and hydrocarbs.

MANAGEMENT OF SMELL

The appearance of smell is probably the most common of the problems associated with the composting process, which luckily can largely be controlled. The factors which determine the anaerobic conditions, therefore, smells are (Dumitrescu 2004, Toledo et al. 2019):

- Excess of humidity.
- Inadequate porosity of the materials for composting.
- Rapid degradation of the sublayer.
- Excessive size of the reactor.

The smell developed in compost can come from the raw materials kept in anaerobic conditions for a long time. Once included in the composting system, these ingredients cause the appearance of smells due to the degradation in anaerobic conditions when a series of compounds like those of reduced sulfur (hydrogen sulfide, dimethyl sulfite, dimethyl disulfite and methyl mercaptan), volatile fatty acids, aromatic compounds and amines are formed. Of all these compounds, ammonia is most widely met, both in the aerobic and anaerobic processes and is also the most monitored. Luckily, it is not a persistent gas because as it is lighter than air (its density is 60% of that of the air) it disperses quickly, distinguishing thus from hydrogen sulfide and other gas.

Ammonia can form by aerobic or anaerobic processes, that is why the control of the composting process is required and it must run aerobically. The appearance of ammonia is due first to the C:N ratio and indicates the fact that nitrogen is in excess compared to carbon (for example when green materials are composted).

In the case of big composting installations, high losses of ammonia can generate the contamination of surface or deep water. Another factor that affects the volatility of ammonia is pH. Gas ammonia and the one in water liquid as ammonia ions are in balance when pH = 9. Should pH exceed this value, the transformation of an additional quantity of ammonia ions into gas ammonia takes place, perceptible by smell (a process which does not happen with an acid pH).

In the case of passively aired systems, which depend on diffusion and natural convection, it is vital to ensure an adequate porosity in order to reduce the resistance to oxygen movement.

Catalyzers which absorb smells can be used. Catalyzers can biodegrade compounds with bad smell through the enzymes involved in the process. The most efficient catalyzers are enzymes that can biodegrade more molecules of smelling organic compounds. They can act both at the surface of the compost and in the atmosphere from the reactor. A series of products of this type are available.

QUALITY CRITERIA OF COMPOST

The fabrication technology and the content of raw materials of compost must meet the EU requirements set up in the Council Order no. 2092/91, annex II., point A, regarding the materials for the soil fertilization and improvement as well as the demands of the International System for Quality Assurance ISO9001. The demands on brown to black, humic fertilizers with loosening, tender structure, the quality requirements and conditions are established as follows:

- When it comes to the size of the granule from the humic fertilizer, at least 60% must pass through the sieve, which is 5 mm in diameter.
- The volume of the humic fertilizer must vary between 0.8–0.9 kg/l.
- The chemical reaction of the material must have a pH of 7.5 ± 0.5.
- Content of dry substance: at least 50%.
- Contents of nutritious elements: N, %/s.u.– at least 0.9; P_2O_5, %/s.u.– at least 0.7; K_2O, %/s.u.– at least 0.5; total soluble salt in water, %/s.u.–1.0 at the most.
- Approximate values: Ca, %/s.u.– 7.9; Mg, %/s.u.– 0.6; Cu, mg/kg/s.u.–31; Zn, mg/kg/s.u.– 180; Mn, mg/kg/s.u.– 355; Fe, mg/kg/s.u.– 7.1; Mo, mg/kg/s.u.– 0.45; Al, mg/kg/s.u.– 7.9; Co, mg/kg/s.u.– 4; B, mg/kg/s.u.– 41.2.
- Toxic elements: Cr, mg/kg/s.u.– 21 at the most; Cd, mg/kg/s.u.– 1 at the most; Ni, mg/kg/s.u.– 20 at the most; As, mg/kg/s.u.– 0.7 at the most; Pb, mg/kg/s.u.– 17 at the most; Hg, mg/kg/s.u.– 0.2 at the most;

- Microbiological data: number of bacteria– 11×107/g; number of actinomycetes– 45×105/g; number of microfungi– 10×105/g.

REFERENCES

Amlinger, F. 1993. Handbuch der Kompostierung. Ludwig-Bolzmann-Institut fur Biologischen Landbau, Wien, Austria.

Amlinger, F., Götz, B., Dreher, P., Geszti, J. and Weissteiner, C. 2003. Nitrogen in biowaste and yard waste compost: Dynamics of mobilisation and availability—A review. European Journal of Soil Biology. 39(3): 107–116.

Blaga, G., Filipov, F., Rusu, I., Udrescu, S. and Vasile. D. (eds) 2005. Pedology. Academic Press, Cluj Napoca, (in Romanian).

Dumitrescu, G.L. 2004. Research on the possibilities for recovery and superior valorisation of household waste by composting. Review of Science and Scientometry Policy, no. special 2005, Grant CNCSIS 602/2004 (in Romanian).

Foereid, B. 2019. Nutrients recovered from organic residues as fertilizers: Challenges to management and research methods. World Journal of Agriculture and Soil Science. 1(4): 1–7. DOI: 10.33552/WJASS.2019.01.000516.

Goyal, S., Dhull, S.K. and Kapoor, K.K. 2005. Chemical and biological changes during composting of different organic wastes and assessment of compost maturity. Bioresource Technology. 96(14): 584–1591. https://doi.org/10.1016/j.biortech.2004.12.012.

He, Y., Xie, K., Xu, P., Huang, X., Gu, W., Zhang, F. and Tang, S. 2013. Evolution of microbial community diversity and enzymatic activity during composting. Research in Microbiology. 164(2): 189–198. https://doi.org/10.1016/j.resmic.2012.11.001.

Pagans, E., Barrena R., Font, X. and Sánchez, A. 2006. Ammonia emissions from the composting of different organic wastes. Dependency on process temperature. Chemosphere. 62(9): 1534–1542. https://doi.org/10.1016/j.chemosphere.2005.06.044.

Ravindran, B., Nguyen, D.D., Chaudhar, D.K., Chang. S.W., Kim, J., Lee S.R., Shin, J.D., Jeon, B.H., Chung, S. and Lee, J. 2019. Influence of biochar on physico-chemical and microbial community during swine manure composting process. Journal of Environmental Management. 232: 592–599. https://doi.org/10.1016/j.jenvman.2018.11.119.

Ruggieri, L., Cadena, E., Martínez-Blanco, J., Gasol, C.M., Rieradevall, J., Gabarrell, X., Gea, T., Sort, X. and Sáncheza, A. 2009. Recovery of organic wastes in the Spanish wine industry. Technical, economic and environmental analyses of the composting process. Journal of Cleaner Production. 17(9): 830–838. https://doi.org/10.1016/j.jclepro.2008.12.005.

Solti, G., Rusu, T., Nagy, M. and Albert, I.O. 2006. Use of Sawdust and Wood Waste for Composting, Ed. Risoprint Cluj-Napoca, p. 98 (in Romanian).

Szajdak, L.W., Gaca, W., Rusu, T., Meysner, T., Styła, K. and Szczepański, M. 2017. Eriophorum-Sphagnum raised bog physicochemical properties of humic acids. *In*: Szajdak, L.W. (ed.), Peat–Physicochemical Properties. LAP Lambert Academic Publishing, pp. 23–42.

Toledo, M., Gutiérrez, M.C., Siles, J.A. and Martín, M.A. 2019. Odor mapping of an urban waste management plant: Chemometric approach and correlation between physico-chemical, respirometric and olfactometric variables. Journal of Cleaner Production. 210: 1098–1108. https://doi.org/10.1016/j.jclepro.2018.11.109.

Vargas-García, M.C., Suárez-Estrella, F., López, M.J. and Moreno, J. 2010. Microbial population dynamics and enzyme activities in composting processes with different starting materials. Waste Management. 30(5): 771–778. https://doi.org/10.1016/j wasman. 2009.12.019.

Villar, I., Alves, D., Garrido, J. and Mato, S. 2016. Evolution of microbial dynamics during the maturation phase of the composting of different types of waste. Waste Management. 54: 83–92. https://doi.org/10.1016/j.wasman.2016.05.011.

Yang, B., Ma, Y. and Xiong, Z. 2019. Effects of different composting strategies on methane, nitrous oxide, and carbon dioxide emissions and nutrient loss during small-scale anaerobic composting. Environmental Science and Pollution Research. 26(1): 446–455.

***CBPA, 2015. Code of good agricultural practices for water protection against nitrate pollution from agricultural sources. Official Gazette of Romania no. 649/27.09.2015, Ministry of Agriculture and Rural Development (in Romanian).

***Order 344/2004 – Order no. 344/2004 for the approval of the Technical Norms regarding the protection of the environment and especially of the soils, when using the sludge in agriculture. Official Gazette of Romania no. 344/708/19.10.2004, Ministry of Agriculture and Rural Development (in Romanian).

***Order 999/2016 – Order no. 999/2016 on the approval of the system of administrative sanctions for cross-compliance applicable to support schemes and measures for farmers as from 2016. Official Gazette of Romania no. 1049/27.12.2016, Ministry of Agriculture and Rural Development (in Romanian).

Chapter 7

Drainage as a Source of Hydrophobic Amino Acids

Lech Wojciech Szajdak
Institute for Agricultural and Forest Environment, Poznań, Poland.

INTRODUCTION

Peatlands are located in every climatic zone occupying >3 million km^2 of the Earth's surface (3% of the Earth's land surface), constituting 70% of natural freshwater wetland. They sequester 0.37 gigatonnes of carbon dioxide (CO_2) a year–storing more carbon than all other vegetation types in the world combined (Lappalainen 1996).

Peatlands are a type of wetlands that are among the most valuable ecosystems on Earth, playing an important role in many biogeochemical cycles. In addition to this, peatlands are critical for preserving global biodiversity, minimizing flood risk and helping to address climate change.

The transformation of vegetable matter to peat accumulation varies widely according to vegetation type, tissue decomposability, root and water table depth, pH, ionic strength, the balance between humification and degradation, climate as well as in relation to carbon and nutrient availability. Reduced litter decomposition in peatlands is supposed to be primarily a consequence of the harsh nature of peatland habitats (low soil pH and temperature and frequent lack of oxygen) and the low nutrient quality of plant litter (Okruszko 1976, Gorham 1991, Johnson and Damman 1993, Belyea 1996, Clymo et al. 1998, Bambalov 2000, Bridgham and Richardson 2003, Bragazza et al. 2007, Gierlach-Hładoń and Szajdak 2010). There are many different ways of classifying wetlands, peatlands and mires depending on the purposes of the classification.

For Correspondence: lech.szajdak@isrl.poznan.pl

The main attributes used in mire and peatland classifications and typologies include floristics, vegetation physiognomy, morphology, hydrology, stratigraphy, chemistry and peat chemical and physical characteristics. Various classifications using one or more of these criteria have also been produced that reflect the objectives of interest groups, for example, conservationists, ecologists, peat extraction industry, forestry and plantation managers.

Two major types of peatlands are bogs (which are mainly rain-fed and nutrient-poor) and fens (mainly fed by surface or groundwater and are more nutrient-rich) (Joosten and Clarke 2002, Szajdak et al. 2011). Both develop under conditions of near-continuous soil saturation that leads to anaerobic conditions, which strongly slows decomposition.

Bogs represent ombrotrophic peatlands with the surface above the surrounding terrain or otherwise isolated from laterally-moving, mineral-rich soil waters. Although some bogs are convex in shape (raised bogs), bogs can also be quite flat or sloping with slight rises at the margin that isolate them from incoming inorganic substances in the water. The main physiognomic groups are open bog and wooded bog (bog forest). There is a large variation of wetness in bogs with a pattern of hummocks and hollows.

Fens characterize minerotrophic peatlands with the water table slightly below, at or just above the surface. Usually, there is slow internal drainage by seepage but sometimes with the over-surface flow. Peat depth is usually greater than 40 cm but sometimes less (for instance adjacent to the mineral edges). Two broad types are *topogenous* (basin) fen and *soligenous* (sloping) fen. The main physiognomic groups of fen are open fen and wooded fen (with *tree cover*, or a sparse tall shrub cover, sometimes called *shrub carr*).

There are over 20 forms of bogs, 19 forms of fens, and 6 forms of swamps that would be considered peatlands (Lappalainen 1996).

Peat is characterized be colloidal behavior and by the irreversible loss of wettability produced by drying. Long-term cultivation and agricultural use of peatlands and their exploitation have revealed a number of effects, including lowering the water table, increased aeration, changes in plant communities and the release of carbon contents. These processes show the disturbance of the thermodynamic balance in peat. The decline in peat soil moisture content resulting from drainage leads to shrinkage of the peat. Volume change due to shrinkage is the result of several forces acting at micro-scale, whereby its mechanism and magnitude differ from those in mineral (clay) soils. It was shown that drainage in particular results in a sharp change of biotic and abiotic conversions and consequent degradation of peat organic matter (Grootjans et al. 1986, Kwak et al. 1986, Lüttig 1986, Bridgham and Richardon 2003, Säurich et al. 2019).

Drainage is the main direct cause of fen habitat degradation, either due to the reclamation of fen or due to the changes in the water flow within fen systems. Lowering of the water level in peatlands activates anaerobic conditions into aerobic ones and accelerates the peat mineralization process. In the first period after drainage, this causes usually an increase of nutrient availability, especially nitrogen and phosphorus, which are released during mineralization (Koeselman and Verhoeven 1995). However, the increased fertility is usually only a short-term effect (Kajak and Okruszko 1990) and therefore additional fertilization is required to sustain economically prospective agricultural production on drained fens, which has a further negative impact on biodiversity. Apart from increasing nutrient availability, drainage also lowers the water storage capacity of peat soils, making them more susceptible to water-table fluctuations and droughts. A further aspect of fen habitat degradation is acidification. This may be related to drainage, which results in partial replacement of groundwater by rainwater (Van Diggelen et al. 1996) but also to the increased atmospheric deposition of nitrogen and sulfur compounds (Gorham 1991, Kotowski 2002).

Drying and wetting of peat soils lead to soil volume changes that result in soil vertical movement and bulk density changes (Schnitzer 1986, Brandyk et al. 2001, 2002).

Up to 35 × 10^6 ha of wetlands has been drained (Bouwman 1990) and it should be noted, drained organic soils are subsiding at the rate of several cms per year (Netherlands: 1.75 cm y^{-1}, Quebec in Canada: 2.07 cm y^{-1}, Everglades in the United States: 3 cm y^{-1}, San Joaquin Delta in the United States: 7.6 cm y^{-1} and Hula Valley in Israel cm y^{-1}) (Terry 1986). Among the reasons for subsidence are shrinkage due to drying, loss of the buoyant force of groundwater, compaction, wind erosion, burning and microbial oxidation, which is a predominant feature of Histosols subsidence. Approximately 73% of the loss of surface elevation in Everglades Histosols is caused by microbial oxidation (Terry 1986). An assumed C release from drained wetlands by oxidation of the organic material of 10 t C ha^{-1} y^{-1} yields a global annual C release of 0.05 to 0.35 Pg C. Drainage of 106 ha of Gleysols causes an extra release of 0.01 Pg C y^{-1}. The resulting global release from Histosols and Gleysols ranges between 0.03 and 0.37 Pg C y^{-1}.

Based on an average storage rate of 200 kg C ha^{-1} in wetlands of the cool climate and assuming an areas of about 350 × 106 ha, the annual accumulation before disturbance has been calculated as 0.06–0.08 Pg C y^{-1} (Armentado and Menges 1986). The total drained in the period 1795–1980 was 8.21910^6 ha for crops, 5.5 × 10^6 ha for pasture, and 9.4 × 10^6 ha for forests respectively.

In the tropics, about 4% of the wetland was reclaimed in the period 1795–1980. The annual shift (loss of sink strength and gain of source strength) in the global C balance is 0.063–0.085 Pg C due to the reclamation of Histosols in cool regions. Including tropical Histosols, the global shift would be 0.15–0.184 Pg C y^{-1} (Bouwman 1990). The potential to increase C levels in soils under cultivation is largely restricted to upland soils. Restoring C sinks in artificially drained wetland soils is unlikely unless they are taken out of agricultural production to natural wetlands (Nieder et al. 2003).

In general, drainage of peat leads to the progressive differentiation of the hydrophobic peptides and total amino acid content in organic matter. In proteins of peats, hydrophobic contacts exist between hydrophobic and hydrophilic structural elements (between the side chains of the radicals of phenylalanine, leucine, isoleucine, valine, proline, methionine and tryptophan). Hydrophobic forces stabilize the tertiary structure of proteins and determine the properties of lipids and biological membranes. The presence of amino acids, hydrocarbon chains and other non-polar fragments in their composition are related to hydrophobic properties of humic substances (Sokołowska et at. 2005). Since organic matter causes soil water repellency to compound, which constitutes a major part of the soil phase of peat and moorsh, it is therefore important to study the influence of chemical soil properties on their wettability. It was observed that the significant changes in chemical properties of transformed organic matter of peat have a significant influence on the sequential changes in physical and hydraulic properties initiated by drainage for agriculture. Due to a high increase of shrinkage, changes in the number of many other properties of peat related to their humic components are observed (Van Dijk, 1971).

Various mechanisms have been postulated in the literature for the characterization of moorsh formations and the degree of peat transformation. Okruszko (1993) proposed the application of a three-point scale for this purpose and divided moorsh into three categories: peaty moorsh (Z_1), humic moorsh (Z_2) and grainy, i.e. proper moorsh (Z_3). Furthermore, on the basis of the character of genetic layers and their thickness three stages of moorshing, process can be distinguished which are Mt I is weak moorshing, Mt II is moderate moorshing and Mt III is strong moorshing. However, it should be noted the above classification schemes are qualitative.

Quantitative approaches have also been developed to classify the degree of the secondary transformation of peats. Usually, such classifications are based on the numerical values of water adsorptivity because the changes in peat mass under the influence of drying are manifested by the decrease of water holding capacity.

Schmidt (1986) suggested a quantitative method for the achievement of the state of moorsh transformation. This method was based on the evaluation of the amount of bound water at the suction power of 100 kP (1992).

Gawlik (1992) proposed to use the index of water adsorptivity (W_1), which expresses the ratio of the water capacity of a sample that has been previously dried at 105°C to its water capacity in the natural state. The values of W_1 lower than 0.36 characterize non-transformed peat fraction. The values of the W_1 index from 0.36 to 0.61 are characteristic for weak and medium forms of transformed peat moorsh. A W_1 index higher than 0.61 and lower than 0.82 characterizes strongly transformed moorshes, while a W_1 index higher than 0.82 pertains to degraded moorshes.

The moorsh soils are very rich in natural organic matter, which consists of a mixture of plant and animal products at various stages of their decomposition and substances synthesized after the breakdown of these compounds (Stevenson 1982). Changes and release of organic matter accompany the process of secondary transformation of peat. In particular, decomposition of plant biomass, transamination of keto acids, root exudates and autolysis of microorganisms release amino acids. These compounds are the main source of soil nitrogen. Amino acids can subsequently undergo mineralization, migration down the soil profile, adsorption and humification (Sörensen 1967, Stevenson 1982).

Soil organic matter takes part in a number of processes coupled with acid-base properties and plays a major role in pH buffering, ionic balance, metal leaching and metal cation concentrations in the soil solution. Various kinds of functional groups have been identified in natural organic matter, including carboxylic-, phenolic- and hydroxylic groups. Nitrogen and sulfur-containing functional groups, such as amino-, amido-, imino-, sulfamino-, thiol-, sulphenic- and sulphonic acid groups can also be present in smaller quantities (Schnitzer 1978, Schnitzer and Khan 1978, Sposito 1989, Stevenson 1994, Swift 1996, Szajdak 2002). In general, organic matter has a negative charge caused by the dissociation of these functional groups. As a result of their varied nature, kind and localization, the organic matter of such groups is highly heterogeneous (Szajdak and Szatyłowicz 2010). The number and acidic strength (dissociation constants) of these groups constitute the primary characteristics of soil organic matter.

From 90 to 95% of the nitrogen in the surface layer of most soils occurs in an organic form. The main identifiable organic nitrogen compounds (30–45%) in soil hydrolysates are amino acids. Amino acids permeate into soils from root exudates and are created in the process of post-harvest residue decomposition. They can also result from the transamination of certain keto acids, compounds that are a source of food for plants and other soil organisms. Amino acids occur in soil in the free and bound fractions; the latter being present in very low concentrations are very unstable and migrate quickly (Życzyńska-Bałoniak and Szajdak 1993, Ryszkowski et al. 1998). Considerable quantities of amino acids occur in soil in the protein fraction bound to humus, mostly to humic acids (Sörensen 1967, Szajdak and Österberg 1996). Humus has a protective effect on the protein complex, preventing its further decomposition (Trojanowski 1973). The protein fraction of humus is included in the organic colloid component of soils (Figure 7.1), where the colloidal

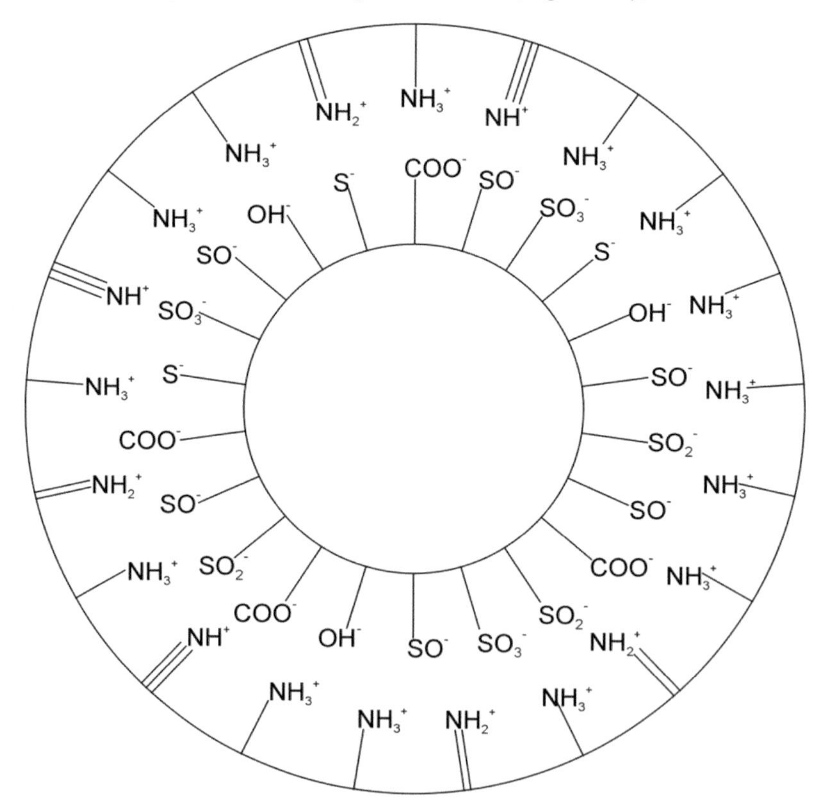

Figure 7.1 Structure of colloidal particle of organic matter.

character of peat soils is stronger than in most mineral ones and the specific surface area of organic colloids is from 2 to 4 times greater than that of montmorillonite minerals (Buckman and Brady 1971, Kwak et al. 1986).

MATERIAL AND METHODS

Samples were taken from 14 sites on peat-moorsh soils in various phases of moorshing process in use as meadows. The sites were located on the low moor of the Wieprz–Krzna canal (Polesie Lubelskie) and in the Biebrza River valley on Kuwasy and Modzelówka peat-bogs. The samples were taken from a depth of 5–10 cm and in strongly peated soils from 5–20 cm. The soil material represented peat soils Z_1 (4 samples) and proper, i.e. granular moorsh Z_3 (10 samples). Indices of secondary transformations were determined according to Gawlik's peat soils method (1996). The differentiation of water holding capacity of peat materials was expressed with the help of the W_1 index.

For analysis of bound amino acids, soil samples were hydrolyzed with 6 M HCl for 24 h at 105°C. Separation and determination of the bound amino acids were carried out on T 339 amino acids analyzer (Mikrotechna-Prague). All the experiments were run in triplicate and the results averaged (Szajdak and Österberg 1996).

RESULTS AND DISCUSSION

In each sample, the number of amino acids determined was up to 26 compounds. The highest total amount of amino acids was found in soil, which belongs to completely degraded peat-moorsh soil. The lowest content of total amino acids (5.30 g·kg) was found in peat moorsh having the lowest value of the W_1 index. During the separation of total amino acids from the studied soils, the following groups were identified: acidic, neutral, basic and sulfur. Results indicated that in all soils neutral amino acids predominated and the content of amino acids ranged from 52.2 to 61.5% of all compounds determined. Moreover, a similar yield of amino acids with a positive net charge at neutral pH and the total amount of acidic amino acids with a negative net charge was found and the proportion of basic amino acids was from 13.2 to 30.0%. Sulfur-containing amino acids in all analyzed samples showed the lowest concentrations. The proportions of these compounds ranged from 10.7 to 21.2%.

Soil amino acids are components of protein and peptides and occur in a stable form. Soil proteins included in organic colloids represent hydrophilic properties. They have a structure denatured to various degrees (Turski et al. 1974). The analyzed soils have differing values of water adsorptivity according to the W_1 index. During studies on the relationship between the W_1 coefficient reflecting, among others, the degree of denaturation of soil colloids, and total amino acids' content in the soil samples and a correlation and linear relationship were found between those two parameters. For the studied soils, this relationship was expressed by the equation $W_1 = 0.0268 \times$ total amino acids' content $+ 0.3537$, characterized by a positive significant value of the directional coefficient and significant positive value of the correlation coefficient (0.93).

Samples were taken from 14 sites on peat muck soils in various phases of the mucking process that are used as meadows. The sites were located on the low moor in the area of the Wieprz-Krzna canal (Polesie Lubelskie) and in the Biebrza River valley on the Kuwasy and Modzelówka peat-bogs. Samples were taken from a depth of 5–10 cm and in very peat-rich soils, ranging from 5–20 cm. The soil material represented peat soils Z_1 (4 samples) and proper, i.e. granular, muck Z_3 (10 samples). Indices of secondary transformation were determined according to Gawlik's method (1992). The soil material, after careful mixing by hand and removing live plant roots, was divided into two parts. The first one (sub-sample A) was soaked for 7 days, put into a special sieve and centrifuged at 1,000 g for one hour. Then it was weighed, oven-dried at 105°C and re-weighed. The same procedure was applied to the second part (sub-sample B) but, in this case, it was air-dried at room temperature before soaking and dried subsequently to remove all water (at 105°C). Differentiation between the water holding capacity of peat materials was expressed with the help of the W_1 index is calculated from the formula:

$$W_1 = \frac{b}{a}$$

where: *a*— centrifugal-moisture equivalent of sub-sample A (in grams of water per 100 g of absolutely dry soil); *b*— centrifugal moisture equivalent of sub-sample B (in grams of water per 100 g of absolutely dry soil).

The lowest content of total amino acids (5.30 g kg^{-1}) occurred in peat moorsh (No 3) which also had the lowest value of the W_1 index. During separation of total amino acids, the following groups were identified: acidic, neutral, basic and sulfur amino acids. Results indicated that neutral amino acids predominated in all soils, and the content of amino acids ranged from 52.2 to 61.5% for all compounds determined. Moreover, similar yields of amino acids with a positive net charge at neutral pH and the total amount of acidic amino acids with a negative net charge were found and the proportion of basic amino acids was shown to be from 13.2 to 30.0% of the total. Sulfur-containing amino acids were presented in the lowest concentrations in all analyzed samples. The proportions of these compounds ranged from 10.7 to 21.2%.

In all soil samples aspartic acid, glycine, alanine, cysteine, 1-methylhistidine and arginine predominated. Particular attention was paid to amino acids having significant importance in soil transformation, i.e. hydroxyproline, proline, β-alanine, lysine, 1-methylhistidine and 3-methylhistidine. Proline, which is transformed from hydroxyproline, occurred in various concentrations from 1.3 to 4.7% of the total amino acids. Hydroxyproline ranged from 0.3 to 3.7%. Proline is a heterocyclic amino acid that undergoes slow decomposition in soil. In acid soils and in the presence of nitrous acid, this compound creates N-nitroso derivatives that are toxic and show cancero-, muta- and teratogenic activity (Figure 7.2) (Kofoed et al. 1981, Larsson et al. 1990).

The greatest amount of this amino acid was found in muck soil from the very strong and secondarily transformed class (soil No 7, 8, 11 and 14) in which the water adsorptivity index ranged from 0.67–0.82. The lowest proline content was found in the soils No 2 (0.09 g kg^{-1}) and No 1 (0.15 g kg^{-1}) with values of W_1 index 0.48 and 0.55, respectively. These belong, therefore, to different classes with respect to their degree of secondary transformation (II and III, respectively).

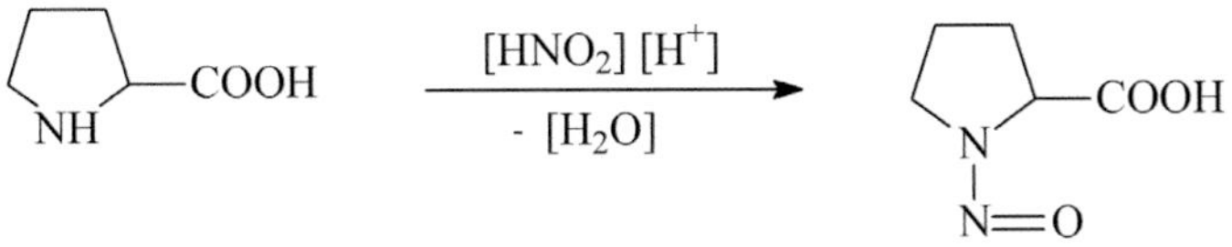

NH —COOH $\xrightarrow[-\,[H_2O]]{[HNO_2]\,[H^+]}$ N(N=O) —COOH

Figure 7.2 Formation of N-nitrozoproline from proline.

β-alanine and lysine, which are components of bacterial cell walls (Stevenson 1972, Durska and Kaszubiak 1980), are indicators of the number or biomass of bacteria in soils (Figure 7.3). In the soil samples, the percentage of lysine (2.9 to 3.9%) was much higher than *β*-alanine (0.4 to 1.8%). The highest total amounts of *β*-alanine and lysine were found in samples 6 and 10 (strongly secondarily transformed), 5 and 6 (very strongly secondarily transformed) and the lowest in 1 and 3.

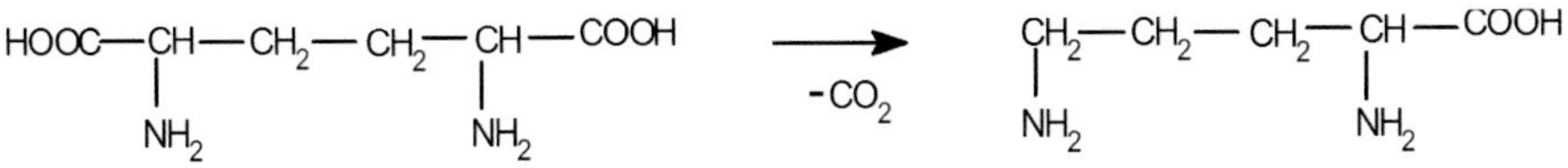

$HOOC{-}CH(NH_2){-}CH_2{-}CH_2{-}CH(NH_2){-}COOH \xrightarrow[-CO_2]{} CH_2(NH_2){-}CH_2{-}CH_2{-}CH(NH_2){-}COOH$

Figure 7.3 Decarboxylation of α,ε— Diaminopimelic acid with the creation of lysine.

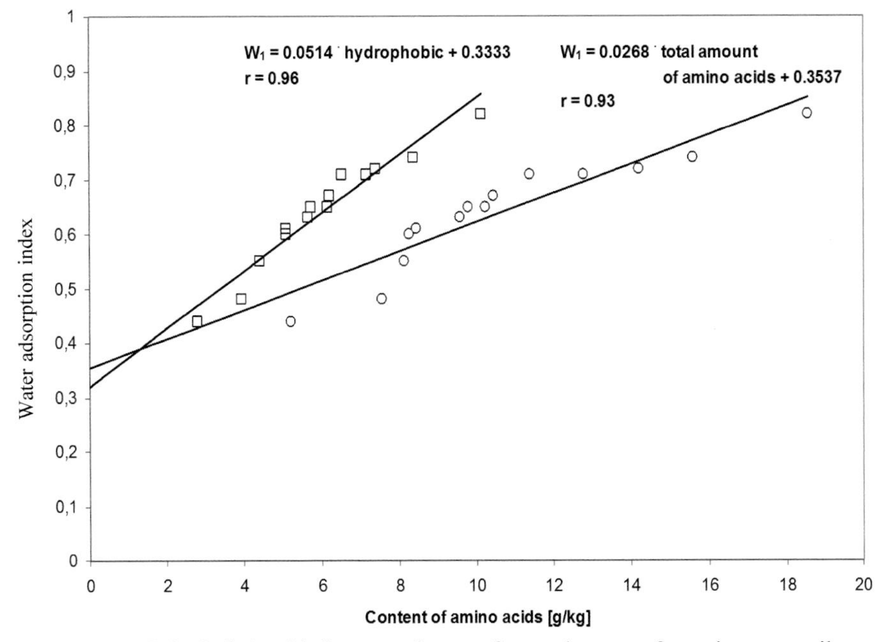

Figure 7.4 Relationship between degree of secondary transformation peat soil W_1 and the content of total and hydrophobic amino acids.

Table 7.1 Total bounded amino acids content in moorsh soil samples ($g \times kg^{-1}$)

Amino Acids	*Moorsh Soil Samples*													
	1	2	3	4	5	6	7	8	9	10	11	12	13	14
	***Ash Content* (% d.m.)**													
	17.56	20.54	22.69	15.14	18.03	37.81	15.80	22.27	20.52	16.26	22.77	18.94	21.24	21.47
	***Bulk Density* ($g \times cm^{-3}$)**													
	0.25	0.28	0.21	0.24	0.36	0.46	0.31	0.39	0.32	0.28	0.30	0.31	0.34	0.29
	***Total Porosity* (vol. %)**													
	84.6	84.7	88.5	85.2	77.8	74.9	80.9	78.7	82.5	82.7	83.6	80.9	81.4	84.1
	Kind of Moorsh Formation													
	Z_1	Z_1	Z_1	Z_1	Z_3	Z_3	Z_3	Z_3	Z_3	Z_3	Z_3	Z_3	Z_3	Z_3
Acidic														
Cysteic acid	0.16	0.05	0.03	0.07	0.19	0.07	0.21	0.45	0.09	0.12	0.13	0.13	0.12	0.13
Aspartic acid	0.49	0.74	0.47	0.92	0.49	0.68	0.72	1.57	0.84	0.83	1.23	0.85	0.64	0.88
Hydroxyproline	0.08	0.48	0.29	0.05	0.28	0.02	0.37	0.43	0.35	0.27	0.29	0.18	0.29	0.58
Threonine	0.43	0.45	0.30	0.54	0.79	0.62	0.75	1.19	0.56	0.55	0.63	0.57	0.44	0.89
Serine	0.40	0.33	0.22	0.44	0.67	0.44	0.66	0.91	0.49	0.49	0.56	0.45	0.38	0.74
Glutamic acid	0.08	0.14	0.14	0.22	0.11	0.08	0.12	0.24	0.09	0.09	0.44	0.07	0.11	0.11
Neutral														
Proline	0.15	0.09	0.20	0.37	0.26	0.28	0.45	0.48	0.35	0.45	0.53	0.27	0.25	0.38
Glycine	0.72	0.76	0.56	1.03	1.11	0.99	1.28	2.15	1.01	1.07	1.38	1.02	0.90	1.37
Alanine	0.46	0.59	0.37	0.66	0.70	0.65	0.78	1.26	0.71	0.71	0.85	0.69	0.67	0.89
α-aminobutyric acid	0.42	0.37	0.21	0.41	0.83	0.57	0.72	1.74	0.57	0.66	0.48	0.54	0.52	0.85
Valine	0.41	0.44	0.27	0.45	0.73	0.57	0.61	0.98	0.52	0.49	0.65	0.55	0.55	0.76
Cysteine	0.94	0.41	0.31	0.68	1.17	0.78	1.09	0.15	0.79	0.72	0.69	0.85	0.57	1.24
Methionine	0.45	0.16	0.10	0.24	0.73	0.12	0.66	1.07	0.50	0.46	0.44	0.55	0.46	0.70
Cystathionine	0.18	0.20	0.12	0.20	0.34	0.28	0.28	0.46	0.24	0.27	0.25	0.27	0.22	0.38
Isoleucine	0.34	0.44	0.33	0.57	0.83	0.58	0.65	0.91	0.53	0.71	0.63	0.63	0.56	0.73

Leucine	0.06	0.09	0.08	0.11	0.10	0.14	0.10	0.15	0.10	0.13	0.11	0.13	0.07	0.16
Tyrosine	0.22	0.30	0.20	0.31	0.45	0.48	0.44	0.65	0.38	0.42	0.40	0.43	0.27	0.59
β-alanine	0.04	0.05	0.04	0.04	0.10	0.17	0.06	0.10	0.03	0.10	0.08	0.06	0.04	0.10
β-aminobutyric acid	–	–	–	0.01	–	–	–	–	–	–	–	0.14	–	0.16
γ-aminobutyric acid	0.02	0.03	0.02	0.02	0.06	0.05	0.05	0.03	0.03	0.03	0.05	0.04	0.02	0.06
Basic														
Ornithine	0.07	0.05	0.01	0.05	0.08	0.04	0.06	0.14	0.05	0.06	0.07	0.07	0.05	0.09
Lysine	0.32	0.29	0.17	0.25	0.40	0.35	0.41	0.72	0.31	0.33	0.38	0.35	0.26	0.51
Histidine	0.10	0.06	0.06	0.09	0.13	0.09	0.14	0.21	0.09	0.08	0.10	0.09	0.07	0.15
1-methylhistidine	0.52	0.11	0.08	0.09	1.29	0.37	0.61	0.67	0.22	0.23	0.09	0.26	0.18	0.92
3-methylhistidine	0.39	0.08	0.04	0.07	1.00	0.29	0.49	0.56	0.17	0.18	0.08	0.19	0.15	0.70
Arginine	0.68	0.84	0.59	0.56	1.37	0.86	1.08	1.33	0.76	0.99	0.84	0.85	0.48	1.51
Total amount	**8.13**	**7.57**	**5.21**	**8.45**	**14.21**	**9.57**	**12.79**	**18.55**	**9.78**	**10.44**	**11.38**	**10.23**	**8.27**	**15.58**
W_1	*0.55*	*0.48*	*0.44*	*0.61*	*0.72*	*0.63*	*0.71*	*0.82*	*0.65*	*0.67*	*0.71*	*0.65*	*0.60*	*0.74*

where: W_1— degree of secondary transformation

Particular attention should be given to the increased total amounts of 1-methylhistidine and 3-methylhistidine in samples 5–8 and 10. The percentages of these substances ranged from 1.6 to 16.1% of the total amount of amino acids in these soils. Since 1-methylhistidine and 3-methylhistidine are products of decomposition of lower plants (mosses, lichens and algae) (Parsons and Tinsley 1975), the high amounts of these compounds in the tested samples indicate higher biomes of these plants in the studied soils (Table 7.1).

Soil amino acids are components of protein conglomerates that occur in a stable form. Soil proteins included in organic colloids are hydrophilic colloids whose structure has been denatured to various degrees (Trojanowski 1973, Turski et al. 1974). The analyzed soils have different values of W_1 (Table 7.1). During studies of the relationship between the W_1 coefficient reflecting, among others, the degree of soil denaturation of colloids, and total amino acids' content in the soil samples, a correlation and linear relationship were found between those two parameters. For the studied soils this relationship is expressed by the equation: $W_1 = 0.0268 \times$ total amino acids content + 0.3537, characterized by a positive significant value of the respective directional and correlation coefficients ($r = 0.93$) (Figure 7.4).

CONCLUSIONS

The moorshing process, as a result of drainage, leads to the progressive differentiation of total amino acids' content. The soils at various stages of moorshing contain various amounts of total amino acids. A relationship, therefore, was found between the total amino acids' content in moorsh and its degree of secondary transformation determined by the index of water adsorptivity (W_1).

REFERENCES

Armentado, T.V. and Menges, E.S. 1986. Patterns of change in the carbon balance of organic soil-wetlands of the temperate zone. Ecology. 74: 755–774.

Bambalov, N. 2000. Regularities of peat soils anthropic evolution. Acta Agrophysica. 26: 179–203.

Belyea, L.R. 1996. Separating the effects of litter quality and microenvironment on decomposition rates in a patterned peatland. Oikos. 77: 529. https://www.jstor.org/stable/3545942?origin=crossref.

Bouwman, A.F. 1990. Exchange of greenhouse gases between terrestrial ecosystems and the atmosphere. *In*: Bouwman, A.F. (ed.), Soils and the Greenhouse Effect. John Willey and Sons, Chichester, pp. 61–129.

Bragazza, L., Siffi, C., Iacumin, P. and Gerdol, R. 2007. Mass loss and nutrient release during litter decay in peatland: The role of microbial adaptability to litter chemistry. Soil Biology and Biochemistry. 39: 257–267. https://linkinghub.elsevier.com/retrieve/pii/S003807170600352X.

Brandyk, T., Szatyłowicz, J., Oleszczuk, R. and Gnatowski, T. 2002. Water-related physical attributes of organic soils. *In*: Parent, L–E. and Ilnicki, P. (eds), Organic Soils and Peat

Materials for Sustainable Agriculture. CRC Press and International Peat Society, Boca Raton, Florida, USA, pp. 33–66.

Brandyk, T., Oleszczuk, T. and Szatyłowicz, J. 2001. Investigation of soil water dynamics in a fen peat-moorsh soil profile. International Peat Journal. 11: 15–24.

Bridgham, S.D. and Richardson, C.J. 2003. Endogenous versus exogenous nutrient control over decomposition and mineralization in North Carolina peatlands. Biogeochemistry. 65: 151–178.

Buckman, H. and Brady, N.N. 1971. Gleba i jej właściwości. PWRiL, Warszawa, pp. 328–333.

Clymo, R.S., Turunen, J. and Tolonen, K. 1998. Carbon Accumulation in Peatland. Oikos. 81: 368. https://www.jstor.org/stable/3547057?origin=crossref.

Durska, G. and Kaszubiak, H. 1980. Occurrence of α-,ε-diaminopimelic acid in soil. II. Usefulness of α-,ε-diaminopimelic acid determination for calculations of the microbial biomass. Polish Ecological Study. 6: 195–199.

Gawlik, J. 1992. Water holding capacity of peat formations as an index of the state of their secondary transformation. Polish Journal of Soil Science. 25(2): 121–126.

Gawlik, J. 1996. Przydatność wskaźnika chłonności wodnej do oceny stanu wtórnego przeobrażenia gleb torfowych. Wiadomości IMUZ, t. 18(3): 197–216.

Gierlach-Hładoń, T. and Szajdak, L. 2010. Physicochemical properties of humic acids isolated from an Eriophorum Sphagnum raised bog. *In*: Kļaviņš, M. (ed.), Mires and Peat. Riga: University of Latvia Press, pp. 143–157.

Gorham, E. 1991. Northern Peatlands: Role in the Carbon Cycle and Probable Responses to Climatic Warming. Ecology Applied. 1: 182–195. DOI.wiley.com/10.2307/1941811.

Grootjans, A.P., Schipper, P.C. and van der Windt, H.J. 1986. Influence of drainage on N-mineralization and vegetation response in wet meadows. II. Cirsio-Molinietum stands. Oecology Plant. 7: 3–14.

Johnson, L. and Damman, A.W. 1993. Decay and its regulation in sphagnum peatlands. Advances in Bryology. 5: 249–296.

Joosten, H. and Clarke, D. 2002. Wise use of Mires and Peatlands–Background and Principles including a Framework for Decision-Making. International Mire Conservation Group and International Peat Society. Saarijärven Offset Oy, Saarijärvi, Finland, p. 304.

Kajak, A. and Okruszko, H. 1990. Grasslands on drained peats in Poland. *In*: Breymeyer, A.I. (ed.), Ecosystems of the World 17A; Managed Grasslands, Elsevier Sc. Publ. Amsterdam, pp. 213–253.

Koeselman, W. and Verhoeven, J.T.A. 1995. Eutrophication of fen ecosystems: External and internal nutrient sources and restoration strategies. *In*: Wheeler, B.D., Show, S.C., Fojt, W.J. and Robertson, R.A. (eds), Restoration of Temperate Wetlands. Willey, Chichester, pp. 91–112.

Kofoed, D., Nemming, O., Brunfeld, K., Nebelin, E. and Thomson, J. 1981. Investigations on the occurrence of nitrosamines in some agricultural products. Acta Agriculture Scandinavica. 31: 40–48.

Kotowski, W. 2002. Fen communities. Ecological mechanisms and conservation strategies. PhD thesis. University of Groningen. Van Denderen. Groningen. pp. 1–181.

Kwak, J.C., Ayub, A.L. and Shepard, J.D. 1986. The role of colloid science in peat dewatering: Principles and dewatering studies. *In*: Fuchsman, C.H. (ed.), Peat and Water. Aspects of Water Retention and Dewatering in Peat. Elsevier Applied Science Publishers, London, pp. 95–118.

Lappalainen, E. 1996. Global Peat Resources. International Peat Society, Geological Survey of Finland, p. 359.

Larsson, B.K., Osterdahl, B.G. and Regner, S. 1990. Polycyclic aromatic hydrocarbons and volatile N-nitrosamines in some dried agricultural products. Swedish Journal Agricultural Research. 20: 49–56.

Lüttig, G. 1986 Plants to peat. *In*: Fuchsman, C.H. (ed.), Peat and Water. Aspects of Water Retention and Dewatering in Peat. Elsevier Applied Science Publishers, London, pp. 9–19.

Nieder, R., Benbi, D.K. and Isermann, K. 2003. Soil organic matter dynamics. *In*: Benbi, D.K., Nieder, R. (eds), Handbook of Processes and Modeling in the Soil-plant System. Food Products Press, The Haworth Reference Press, Imprints of the Haworth Press. Inc., New York, pp. 345–408.

Okruszko, H. 1976. The principles of the identification and classification of hydrogenic soils according to the need of reclamation. Biblioteka Wiadomości IMUZ. 52: 7–53.

Okruszko, H. 1993. Transformation of fen-peat soils under the impact of draining. Zeszyty Problemowe Postępów Nauk Rolniczych. 406: 3–73.

Parsons, J.W. and Tinsley, J. 1975. Nitrogenous substances. *In*: Gieseking, J.E. (ed.), Soil Components, Vol. 1. Organic Components. Springer Verlag, Berlin, pp. 263–304.

Ryszkowski, L., Szajdak, L. and Karg, J. 1998. Effects of continuous cropping of rye on soil biota and biochemistry. Criticical Review Plant Science. 17(2): 225–244.

Säurich, A., Tiemayer, B., Don, A., Fiedler, S., Bechtold, M., Wulf Amelung, W., and Freibauer, A. 2019. Drained organic soils under agriculture—The more degraded the soil the higher the specific basal respiration. Geoderma, 355: 11391. https://doi.org/10.1016/j.geoderma.2019.113911

Schmidt, W. 1986. Zur Bestimmung der Einheitswasserzahl von Torfen. Arch. F. Acker–Pflanzenbau Bodenkunde. 30(5): 251–257.

Schnitzer, M. 1986. Water retention by humic substances. *In*: Fuchsman, C.H. (ed.), Peat and Water. Aspects of Water Retention and Dewatering in Peat. Elsevier Applied Science Publishers, London, pp. 159–176.

Schnitzer, M. and Khan, S.U. 1978. Soil Organic Matter. Elsevier Scientific Publishing Company, Amsterdam, pp. 261–262.

Schnitzer, M. 1978. Chapter 1 Humic Substances: Chemistry and Reactions. pp. 1–64. Available from: https://linkinghub.elsevier.com/retrieve/pii/S0166248108700163.

Sokołowska, Z., Szajdak, L. and Matyka-Sarzyńska, D. 2005. Impact of the degree of secondary transformation on acid-base properties of organic compounds in mucks. Geoderma. 127: 80–90.

Sörensen, J.H. 1967. Duration of amino acids metabolites formed in soil during decomposition reactions. Soil Science. 104: 204–241.

Sposito, G. 1989. The Chemistry of Soil. Oxford University Press, New York, pp. 24–98.

Stevenson, G. 1972. The Biology of Fungi, Bacteria and Viruses. PWRiL, Warszawa. p. 280 (in Polish).

Stevenson, F.J. 1994. Humus Chemistry: Genesis, Composition, Reactions, 2nd Ed. Wiley and Sons, New York, p. 512.

Swift, R.S. 1996. Organic matter characterization. *In*: Methods of Soil Analysis. Part 3. Chemical Methods-SSSA Book Series No. 5, Madison. WI, 1011–1069.

Szajdak, L. and Österberg, R. 1996. Amino acids present in humic acid from soil under different cultivations. Environment International. 22(3): 331–334.

Szajdak, L. 2002. Chemical properties of peat. *In*: Ilnicki, P. (ed.), Peat and Peatlands. Wydawnictwo Akademii Rolniczej im. A. Cieszkowskiego, Poznań, pp. 432–450 (in Polish).

Szajdak, L. and Szatyłowicz, J. 2010. Impact of drainage on hydrophobicity of fen peat-peat-morsh soils. *In*: Kļavinš, M. (ed.), Mires and Peat. Riga: University of Latvia Press, pp. 158–174.

Szajdak, L.W., Szatylowicz, J. and Kõlli, R. 2011. Peat and peatlands, physical properties. *In*: Gliński, J., Horabik, J. and Lipiec, J. (eds), Encyclopedia of Agrophysics. Encyclopedia of Earth Sciences Series, Springer, pp. 551–555.

Terry, R.E. 1986. Nitrogen transformations in Histosols. *In*: Chen, Y. and Avnimelech, Y. (eds), The Role of Organic Matter in Modern Agriculture. Martinus NIjhoff Publishers, Netherlands, pp. 55–69.

Trojanowski, J. 1973. Przemiany Substancji Organicznych w Glebie. PWRiL, Warszawa, pp. 67–70 (in Polish).

Turski, R., Domżał, H. and Słowińska-Jurkiewicz, A. 1974. Wpływ frakcji koloidalnej z uwzględnieniem próchnicy na maksymalną higroskopijność, granice konsystencji i pęcznienia gleb lessowych. Roczniki Gleboznawcze. 25(3): 85–99.

Van Diggelen, R., Molenaar, W.J. and Kooijman, A.M. 1996. Vegetation succession in a floating mire in reaction to management and hydrology. Journal of Vegetation Science. 7: 809–820.

Van Dijk, H. 1971. Colloid chemical properties of humic matter. *In*: McLaren, A.D. and Skujins, J. (eds), Soil Biochemistry. Marcel Dekker, New York, p. 21.

Życzyńska-Bałoniak, I. and Szajdak, L. 1993. The content of bound amino acids in the soil under rye monoculture and Norfolk crop rotation in different periods of plants development. Polish Journal of Soil Science. 26(2): 111–117.

Chapter 8

Nitrogen Status of Estonian Agricultural Landscapes' Soil Cover

Raimo Kõlli*, Karin Kauer and Tõnu Tõnutare

Estonian University of Life Sciences, Institute of Agricultural and Environmental Sciences, Tartu, Estonia.

INTRODUCTION

The nitrogen plays a leading role in the formation and functioning of whichever terrestrial ecosystem. From the nitrogen status (NS) of soil cover (SC) depends on area's potential productivity, functioning activity, its utilization (application) mode (or land use) and suitability for crops. NS of soils depends on various soil properties, land use and ambient area ecological conditions (Garten and Ashwood 2002, Xue et al. 2013, Marty et al. 2017). SC is a multifunctional basis for the landscape (Kõlli and Ellermäe 2001, Arold 2005).

During long-lasting agricultural activity for the agricultural areas in Estonia has been taken the best as possible soils of the region (Kask 1975, Mander et al. 1995, Kõlli and Tamm 2012). In the Northern part of Estonia for the best soils on agricultural landscapes are typical leached and eluviated soils, but in the Southern part, it is pseudopodzolic and ameliorated sod-podzolic soils (Kokk and Rooma 1978, Kokk 1995, Kõlli and Ellermäe 2003). In all agriculturally used territory, these soils are accompanied in more or less extent by the drained gley-soils and peat soils, which are bordered on their transitional area by the peaty soils

*For Correspondence: raimo.kolli@emu.ee

(Kõlli et al. 2009a). For the essential diversifiers of agricultural landscapes are the synlithogenic (Fridland 1982) or abnormally developed soils for which are formed in specific soil-forming conditions erosion-affected, alluvial and coastal soils (Kokk 1977, Kauer et al. 2004, Kõlli et al. 2010b).

The distribution pattern, appearance and functioning of the landscape depend besides the soils' properties and diversity also on the extra-soil natural agents such as climatic conditions, geology and hydrography of the area (DeBusk et al. 2001, Shaffer and Ma 2001). Agricultural landscape condition and usefulness for society depend very much as well on the policy of application local natural land resources, for the basis of this is the kind of land use and the intensity of land management. By the Estonian Land Board data, on December 31, 2018, the area of agriculturally-used land was ca 1.284 10^6 hectares from which the arable lands formed 81.5% but the grasslands were 18.5% (Estonian Land Board 2018). Therefore, under agricultural activity was taken ca 30.3% from the whole Estonian soil cover (Kõlli et al. 2009b).

In the characterization of landscapes fabric and functioning, the authors of this work are preferred the ecosystem approach (Kõlli and Kanal 2010). In this case of approach is essential in studying landscape functioning capability to take into account besides SC properties as well the character of plant cover, the assemblages of organisms adapted to soil condition and the hydrological regimes of ambient territory. The best indicator for characterizing agricultural landscape's functioning intensity is the potential fertility of SC, which is based on the nutritional and humus status of all soil species presented in the SC. In this work, the main attention was paid to the NS of soils. It is important to mention that most of the nitrogen and certain part of other nutritional elements are cycling in the composition of organic matter. Therefore, the nutritional and NS of the ecosystem are determined partly as well by the humus status of the soil.

The main quantitative indexes of organic matter flux via the landscape are the annual phytomass productivity and connected with this annual litterfall intensity. Both of these indexes are closely connected with organic carbon (OC) and total nitrogen (NT) flux and cycling in the ecosystem (Chapman et al. 2013). In the case of organic matter decomposition in soil, the captured in it nitrogen and other elements are switched into the new cycle of elements' turnover or the production process of ecosystems. Also, the denitrification processes in various soils are in a great extent influenced by their NT content (Szajdak and Gaca 2010). For understanding these processes is important to know the NT and soil OC sequestration capacities of different soil types in diverse land use conditions and as well the distribution of NT and coupled with it OC, i.e., the ratio C:N by separate soil layers.

From the thickly coupled elements pair OC and NT have been considered completely from the aspect of biological turnover that studied the fluxes of OC. As a result of this, there are available relatively variegated data on OC accumulation in a dominated Estonian forest, arable and grasslands' soil covers as well formed on the different kinds of ecosystems' plant cover (Kõlli et al. 2010c, Köster and Kõlli 2013). At the same time, the analogs researches on NT stocks and annual fluxes are much modestly carried out and the data on soils NS are relatively few in numbers. The gaps in available data enfold NT concentration and stocks of different soil types and land use conditions; relationships of NT with OC and their common bearer–soil organic matter; changes of NS indices in soil profiles in vertical plan and in the case of land-use change. These have been very modestly been generalized as well as the NS of subsoils and parent materials.

As in Estonia the sequestration of OC into SC and its role in functioning ecosystems have been studied relatively versatilely and long-lastingly (Kokk and Rooma 1983, Kõlli et al. 2010b), the certain elaborations of our previous researches may be taken as the basis for the current study. These works are for good prerequisite in studying of both arable (Kõlli and Ellermäe 2003, Kõlli et al. 2010b, Suuster et al. 2012) and grasslands' soils (Kauer et al. 2004, Kõlli et al. 2007). The main tasks of the actual study, which based on various available for us databases are:

1. To present the statistically elaborated data about NT concentration (g kg^{-1}) and superficial densities (Mg ha^{-1}) by different soil layers of dominated arable and grassland soil species.
2. To distinguish the correlation and interaction of NT with different alternative parameters, which reflect the soil NS as (a) the relationships of NT with OC on the basis of index C:N and (b) the correlation of NT contents with different soil ecological properties using for this as index the correlation coefficients.
3. To explain the role of soils' NT in the forming and functioning of landscapes by different pedo-ecological groups of soils with explaining general regularities in the forming of soils' NS and humus cover types.
4. To generalize the data provided at the level of soil species and/or varieties by soil pedo-ecological groups as normal or post-lithogenic and abnormal or synlithogenic mineral and organic soils by land use and landscape types. By Fridland (1982), the post-lithogenic mineral soils are not influenced by sediment accumulation transported aside or elimination of soil material from a superficial layer, i.e., they are formed in normal soil-forming conditions.

TERMINOLOGY AND METHODOLOGY

Terminology

The NS (nitrogen status) of soil is in principle the character of soil's NT management or NT throughout flux via the SC. It begins with organic matter and sequestrated in it NT influx into the soil, followed by variegated mutual processes with soil living, liquid, gaseous and solid phases until to its sequestration in the stabilized form in soil organic matter (SOM) or its mineralization and elimination from the soil. The main parameters of soils' NS characterization are the concentration (g kg^{-1}) and stock density (Mg ha^{-1}) of NT in different soil layers. Secondary or supporting parameters of soils' NS characterization are ratio C:N and annual NT turnover rate (Mg ha^{-1} yr^{-1}). In actual work mainly the data about C:N ratio is treated.

In the quantitative characterization of landscape's NS the term 'SC' (soil cover) or solum is in use. SC embraces the superficial landscape layer influenced by soil-forming process and consists of humipedon (HP) and subsoil (SS). The SC thicknesses (depth from the soil surface to unchanged parent material) depend to a great extent on SC moisture conditions, calcareousness and texture, reaching in automorphic pseudopodzolic to 100–110 cm, but the thickness of permanently wet gley-soils is in most cases in the limits 40–65 cm. For the benchmark thicknesses of peat soils' (or histosols) SC in the current study 50 cm is taken.

HP (humipedon), named by Zanella et al. (2011, 2018), known as well as epipedon and humus cover, encompasses the most active superficial (topsoil) part of SC via which the dominant share of SOM and sequestrated into it NT cycling takes place. The HP consists of humus, raw-humus or peat horizons (accordingly A, AT and T) and is closely coupled with plant cover. The AT-horizon is a transitional horizon between humus and peat horizons. The AT-horizon is determined on the basis of SOM or SOC content for which percentages are in the limits of 7–35 and 4–20, respectively (Astover et al. 2013). The AT- and T-horizons of mineral soils were always formed in epigleyic soils. For the benchmark thicknesses of peat soils' HP in actual work, 30 cm is taken.

SS (subsoil), which underlies biologically active HP, consists in the case of mineral soils from the illuvial (B) or sequence eluvial-illuvial (E–B) horizons. In peat soils, the SS embraces a peat layer located in the depth from 30 to 50 cm. Therefore its thickness is 20 cm.

Metric soil layer (ML) enfolds all soil horizons located in the layer from the soil surface to the depth of one meter. It consists of the SC and from underlaid it parent material (PM) or C and/or ½BC, Cg and CG horizons, which do not belong to the SC and are by their essence a ground material. Sequestrated into parent material NT is rather buried stock as it does not take part in nitrogen cycling between plant and soil.

PM (parent material) or substrate underlies the SC, whereas its thickness enfolds in actual calculations the soil layer from the bottom of SC to the depth of one meter.

The different soil layers are related with others as belonging to the greater subsets,

$$\mathrm{HP} \subset \mathrm{SC} \subset \mathrm{ML} \tag{1}$$

and (2) as the intersections

$$\mathrm{HP} \cap \mathrm{SC} \cap \mathrm{ML} \tag{2}$$

In the calculation of SS and PM volumes and contained in the NT stocks, the following formulas were used:

$$\mathrm{SS} = \mathrm{SC} - \mathrm{HC} \tag{3}$$

$$\mathrm{PM} = \mathrm{ML} - \mathrm{SC} \tag{4}$$

In the assessment of quantitative data by different soils, the following Estonian soil classification (ESC) taxa were used: (1) soil species, which is the taxon of ESC identified by soil-forming processes and (2) soil variety, which is the subdivision of soil species, identified by soil texture in the case of mineral soils and by peat decomposition stage in the case of peat soil species.

For the qualitative characterization of the soils' humus status in Estonia, the term humus cover type (*pro* humus form) and their classification are in use. Humus cover type, which characterizes (reflects) SOM formation ecology, is a good index as well for evaluation of NS at the landscape level. Totally in Estonian humus cover classification, 27 types have been separated for natural soils and 13 types for arable soils (Kõlli 2018).

Methodological Principles of Field Works, Data Calculation and Laboratory Analyzes

For explaining the NT role in the development and functioning of agricultural landscapes, the pedocentric approach is used (Kõlli and Kanal 2010). For the basis of this approach are the data about soil species (distribution, morphology, content of NT and OC, ecological or soil-forming conditions) and regional peculiarities of formed soil associations (Estonian Land Board 2012). The pedodiversity of the landscape depends directly on soil-forming conditions of the area among them on soil parent material and its deposition character (relief). Shortly saying–the pedodiversity depends on areal geodiversity.

The NT, OC and different agrochemical indexes stocks or superficial densities in different soil horizons were at first calculated on the basis of their concentrations and bulk densities in corresponding soil horizons. Furthermore, the stocks received by horizons were summarized by HP, SS and PM layers, which were for the basis in presentation densities of SC and ML. All laboratory analyzes were done from the fine-earth, which was composed of the particles with a diameter <1 mm and was separated by sieving. In all calculations, the content of coarse fractions, which was determined in field research, was taken into account. The mean weighted NT concentration of SS and PM (calculated in relation to fine-earth and whole soil mass) were used as ecological indicators for comparable analysis of different soil species and/or ecological groups. It should be mentioned that in most cases the HP of arable soils consists mostly of one horizon i.e., of humus or A horizon.

Total nitrogen content was determined by the Kjeldahl method with digestion followed by distillation, and titration with alkali (Bremner 1960).

The content of OC was determined by wet combustion method of Tjurin, where humified SOM is oxidized by 0.2 M solution of potassium dichromate with sulfuric acid by external heating and excess dichromate is determined by titration with ammonium ferrous sulfate (Tjurin 1935, Arinoushkina 1970, Orlov and Grišina 1981).

The particle size distribution is done by Kachinsky (1965). The volume of coarse soil fractions (ø >1.0 cm) was determined during the field research (Astover et al. 2013). In the laboratory, the content of fine-earth (ø <1.0 mm) and of fraction 1–10 mm in soil samples was determined by sieving, but the particle size distribution of fine-earth by using the pipette method. The basis of Kachinsky method is the content of physical clay (ø <0.01 mm) by which the soil textural classes are determined. The used in simplified texture formula symbols (abbreviations) are followings: kr– fine-gravel, l– sand, sl– loamy sand, ls (ls1 and ls2)– loam (i.e., light and medium loam or sandy loam and loam), s (ls3 and s1–2)– clay (clay loam and clay), p– massive limestone, r– ryhk (angular fraction of calcareous origin), v– pebble (rounded calcareous fraction) and t– peat (t3– well and t2– moderately decomposed). For the determination of coarse fractions (r and v) content stages (in % in fine-earth) to their symbols the numbers (1– 2–10%, 2– 10–20%, 3– 20–30% and 4– 40–50%) are added. Therefore, the textural classes of Kachinsky and WRB are not matched in one to one way, but in general line, they should be well understood to the soil scientists.

The pH_{KCl} was determined in 1 M KCl solution in relation to soil/solution with a ratio of 1:2.5, and materials from peat horizons at a ratio of 1:5 electrometrically by means of pH-meters whereas the used method is adequate to ISO 10390. Alternatively, to the data of soil horizons' pH the exchangeable acidity (EA) and hydrolytic acidity (HA) concentration and stocks were used. The EA (exchangeable H^+) and Al (Al^{+++}) were extracted with 1 M KCl solution that afterwards separated titration of total exchangeable acidity ($H^+ + Al^{+++}$) and H^+ using the Sokolov method. The HA and basic cations (BC) were extracted with 1 M $NaCH_3COO$ and determined by titration or by the Kappen method (Vorobyova 1998). The cation exchange capacity (CEC) was calculated by summing BC and HA, but the base saturation stage (BSS) using the formula: BSS = BC/CEC × 100. The quantitative evaluation of acidity per SC layer (EA and HA in kmol ha^{-1}) was performed using its concentration (in mmol kg^{-1}) and the volume weight of the layer.

For all statistical analyzes, the STATISTICA version 13.4 (StatSoft Inc.) software package was used. One-way ANOVA was applied to test the effect of soil species on NS indices. The relationships between different soil agrochemical parameters were tested by using Spearman correlation analysis. The level of statistical significance was set at $P < 0.05$.

PEDO-ECOLOGICAL CONDITIONS AND USED DATA

General Pedo-ecological Characterization of Estonian Agricultural Landscape

The SC of Estonian landscapes is typical to North-eastern Europe. Locating on the transition area between continental and marine climates in mild and wet pedo-climatic conditions of the temperate zone, in natural areas of Estonia, the coniferous, mixed (coniferous-deciduous) and deciduous forests are developed (Laasimer 1965, Raukas 1995). Presently, the forests occupy approximately half of the whole Estonian SC. The main part of parent materials of Estonian soils is derived from the glacial and aquaglacial *Quaternary* deposits. For the parent material of half mineral soils are *Pleistocene* tills. The re-worked from tills glaciofluvial, glaciolacustrine, alluvial and aeolian sediments are distributed alternatively with tills (Raukas and Teedumäe 1997). By the intensified agricultural activity during the last two centuries, the most productive areas of Estonia (soils suitable for crop cultivation and grasslands) have been turned into arable, pastured or hay-lands (Mander et al. 1995, Raukas 1995, Arold 2005). Based on Estonian Weather Service the Estonian annual average air temperature is +6.0°C and annual precipitation varies between 578 and 660 mm https://www.ilmateenistus.ee/2015/01/uued-kliimanormid-1981-2010/.

The present research is limited by agriculturally used soils, which form ca 30.3% of the total Estonian soil cover that is ca 4.240 10^6 ha (Kõlli et al. 2009b). According to the moisture conditions, the greatest share (ca 46%) of the agriculturally used soils form automorphic soils. The share of moist (*endogleyic*) soils is ca 23%, wet mineral *(epigleyic)* soils ca 22% and peat soils 9%. With the best plant growing potential loamy soil form ca 44% of agriculturally used soils. The shares of sands, loamy sands, clays and peats in the agriculturally used soil cover are accordingly 12, 28, 6 and 9%.

DATABASES AND DATASETS

The quantitative data on arable and grassland soils' NT content, agrochemical and ecological characteristics, treatments about relationships of NT with soil OC as well information on SC geographical peculiarities are used in the study that originates from different databases (DB).

DB Monitoring Areas of Arable Soils (MNA)

The research areas (RA) for forming DB MNA were founded in 1983–1987 by Est-Agriproject (Lehtveer and Kokk 1995). For monitoring dominating arable soil species and/or varieties total of 13 RA with 79 plots were established. Totally, it fixed ca 40 soil stable parameters of the soil species found on different plots. The describing of plots' pedo-ecological conditions was followed by studying the dynamics of ca 22 soil parameters. The used in this work data on NS and of background pedo-ecological conditions originated from the years 1983–1994. The DB MNA enfolds 2–4 watching periods of each plot and consists of two following data sets (DS):

DS MNA–1 enfolds 12 RA with 70 plots and 14 soil species. By these data were characterized in the work the main parameters of soils' NS (NT concentration, NT stock and C:N ratio), main arguments of soil NS (OC content and connected tightly with NS soil species properties) and background data in relation to HP by different soil species of normally developed mineral soil species (Table 8.1).

DS MNA–2 was established on the influenced by moderate water erosion area. On this RA with 9 plots, the soil properties and their development were monitored in relation to HP of 16 soil species. Twelve soil species found on this influenced by erosion area may be classified as abnormally developed soils, but four of them as normally developed soils.

DB Pedon

The soil profile horizons DB Pedon was created by us during 1967–1985 but was afterward (in 1986–1995 and 1999–2002) updated and revised (Kõlli et al. 2009b). The DB Pedon contains data on 211 forest-, 160 crop- and 90 grassland soils' profiles. The RA established for studying arable and grasslands' soil profiles are located in a scattered form on the territory of Estonia. If on all RA-s the separate soil horizons samples were taken and analyzed practically and totally, then the bulk density samples were taken from approximately one-tenth of profiles. In most cases, the bulk density of different soil horizons was calculated (ascertained) by us through indirect way, i.e., using the generalizations of previously received bulk density data, which were systematized by soil species and their genetic horizons. Besides that, the pedotransfer functions for soil horizons bulk density calculations were also used (Suuster et al. 2011).

DS Pedon–3 is an excerpt from the DB Pedon, which contains data on more than 160 arable soil profiles given by individual RA. Totally, the DS Pedon–3 enfolds more than 25 soil species data, therefore enfolding most of the dominating types of Estonian arable soils.

On the basis of Pedon–3 the tasks 1 and 2, and partially also the tasks 3 and 4, are treated. Tasks 3 and 4 explain NT relation with OC with a large number of ecological characteristics and with main arguments in soil functioning and land management. On the basis of initial data (given by soil horizons) in actual work, the studied parameters are given by three soil profile layers (HP, SC and ML) and their composing parts (SS and PM) for normally developed mineral soils (Table 8.1).

DS Pedon–4 is also an excerpt from the DB Pedon, but it contains the grassland soil profiles' data, given by RA. Totally, Pedon–4 enfolds gathered from 90 RA soil properties data, representing all dominating grassland soil types of Estonia. With DS Pedon–4 was resolved analogs to arable soil tasks (1–4) but paying much more attention besides normally developed peaty and peat soils as well to abnormally developed alluvial and coastal soils (Table 8.1).

DB Estonian Soils in Numbers (ESN)

Estonian soils in Numbers (ESN) were formed as a result of large scale soil mapping of Estonia (Estonian Land Board 2012). The DB ESN contains, among others, the statistically analyzed data on arable soils' agrochemical properties, presented by soil diagnostic horizons and soil species or/and varieties (Kokk and Rooma 1978, 1983).

DS ESN–5 used in this work is an excerpt from the DB ESN for demonstration geographical differences of one soil species (pseudopodzolic soil) in NS and its background chemical characteristics (Teras and Rooma 1985, Rooma 1987).

DB on Grassland Soils (GRS)

Grassland soils (GRS) was formed by T. Köster and K. Kauer in 2001–2005. The *DS GRS–6* from the DB GRS enfolds four coastal soil species sampled totally from 26 RA-s (Kauer et al. 2004).

DB on Forest Soil Properties (FSP)

Forest soil properties (FSP) formed by E. Asi research group in the framework of the project Biosoil. *DS FSP–7* is an excerpt from DB FSP, which is used in actual work for characterization of deep fen soils SS and PM layers' NT content, OC content and C:N ratio formed in natural conditions (Kõlli et al. 2010a).

DISTRIBUTION OF ARABLE AND GRASSLANDS' SOIL SPECIES IN SOIL COVER OF ESTONIAN AGRICULTURAL LANDSCAPE

The pedological characterization on each soil species (denominated by ESC code and name) includes essential information on soil properties as moisture regime, soil texture, calcareousness/acidity, the fabric of HP and others. The list of studied soil species' codes and names by soil groups is given in Table 8.1, whereas the erosion-affected soils will be presented hereafter in separate Table 8.2

The soil names of national databases were converted into reference soils and qualifier system of World Reference Base (WRB) for soil classification (Estonian Land Board 2012, IUSS WG WRB 2015). This conversion with each soil position in relation to moisture and litho-genetical scalars of normally developed mineral and peaty soils' composite matrix (Figure 8.1) is presented in Table 8.2.

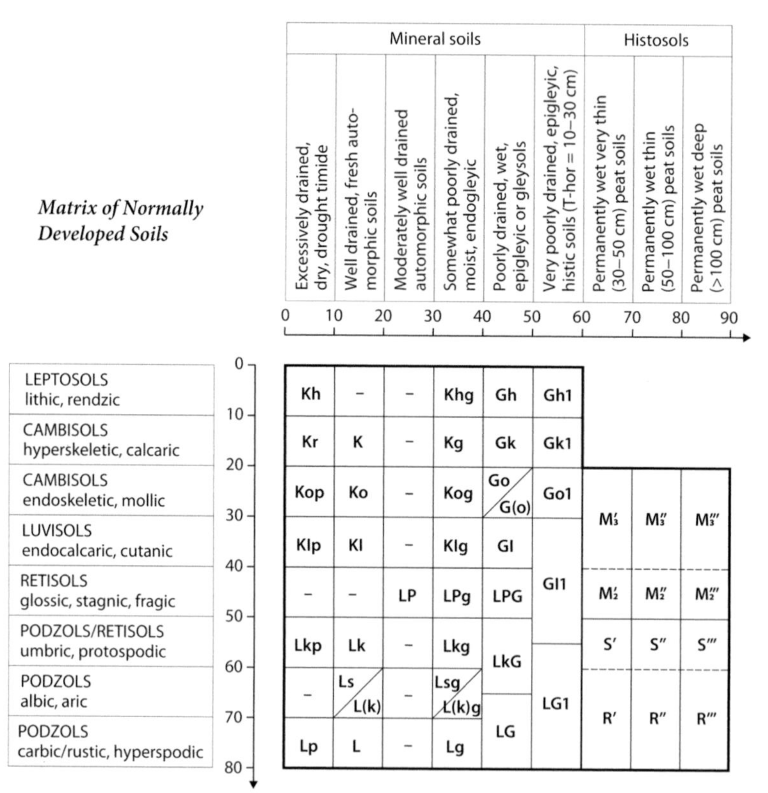

Figure 8.1 Matrix of normally developed soils with soil codes of Estonian Soil Classification. On the horizontal scalar the soils moisture conditions, but on vertical, i.e., litho-genetical scalar, the correlation with WRB reference soils are given. For the soil names after their codes see Table 8.1.

Table 8.1 The codes and names of soil species by Estonian soil classification: sampling, areal distribution and dominating texture

Code	*Name* — *By Estonian Soil Classification*	*Sampling*[1] *MNA/PDN*	*Percentage from Arl/Grl*[2] *Area*	*Percentage from Agricultural Land*	*Texture of Soil Profile*[3]
	I. Normally Developed Mineral Soil Species				
Kh	Limestone rendzinas	0/11	0.8/1.9	1.0	$r_2$ls/p
Kr	Pebble-rich rendzinas	130/8	3.1/0.7	2.7	$r_3$ls/r
K	Pebble rendzinas	1449/15	5.9/0.6	4.9	$r_2$ls/$r_4$ls
Ko	Typical leached soils	664/25	9.7/4.6	8.8	ls/$r_2$ls
KI	Eluviated soils	943/4	6.3/4.2	5.9	ls/$r_1$ls
LP	Pseudopodzolic soils	1100/13	15.2/2.0	12.8	sl/ls
Lk	Sod-podzolic soils	32/21	3.3/3.1	3.3	l, krl, sl
Khg	Gleyed limestone rendzinas	0/1	0 /4.9	0.9	$r_2$ls/p
Kg	Gleyed pebble rendzinas	34/7	2.1/1.5	2.0	$r_2$ls/$r_4$ls
Kog	Gleyed typical leached soils	312/8	4.8/7.0	5.2	ls/$r_2$ls
KIg	Gleyed eluviated soils	497/4	7.2/7.1	7.2	ls/$r_1$ls
LPg	Gleyed pseudopodzolic soils	401/0	6.2/1.9	5.4	sl/ls
Lkg	Gleyed sod-podzolic soils	0/7	1.9/4.1	2.3	l, sl, ls
Gk	Pebble gley-rendzinas	262/4	0.9/4.3	1.5	$r_2$ls/$r_4$ls
Go	Leached gley-soils	203/7	4.6/3.2	4.3	ls/$r_2$ls
G(o)	Saturated gley-soils	0/12	4.8/3.3	4.5	ls, sl, s
GI	Eluviated gley-soils	288/6	5.8/3.2	5.3	ls, sl
LkG	Sod-podzolic gley-soils	216/2	0.7/2.7	1.1	l, sl, l/ls
	II. Normally Developed Peaty Soils				
Go1	Saturated peaty gley-soils	0/2	1.3/5.0	2.0	t_3/ls, sl
GI1	Unsaturated peaty gley-soils	0/2	1.2/4.8	1.9	t_2/l, sl
	III. Normally Developed Peat Soils				
M3	Well decomposed fen soils	0/9	5.5/6.7	5.7	t_3
M2	Moderately decomposed fen soils	0/4	2.2/3.1	2.4	t_2
	IV. Alluvial Soils				
Ag	Gleyed alluvial soils	0/1	0.05/1.2	0.3	sl, l
AG	Alluvial gley-soils	0/3	0.1/3.5	0.7	sl, l
AG1	Peaty alluvial soils	0/4	0.05/1.0	0.2	t_2/sl, l
AM	Alluvial fen soils	0/8	0.6/3.2	1.1	t_2
	V. Coastal Soils[4]				
Gr	Coastal gley-soils	0/26	0/1.7	0.3	sl, ls, l
Gr1	Peaty coastal soils	0/5	0/0.8	0.1	t_2/sl, ls
ArG	Coastal alluvial gley-soils	0/5	0/0.8	0.1	sl, ls, l
	VI. Erosion-Affected Soils				
E-D	Association of EAS[5] (see Table 8.3)	188/0	5.6/7.9	6.0	–

[1]Sampling: MNA— total number of determinations during monitoring period by different plots, and PDN— number of RA-s; [2]Land use: Arl— arable land, Grl— grassland; [3]Simplified formula of soil profile texture; for symbols of soil texture see pages 167–168; [4]Data originated from *DS GRS–6*; [5]EAS— erosion-affected soils.

The matrix of normally developed soils (as a pedo-ecological background) is used in actual work for two purposes 1) in the characterization of soil-forming conditions and 2) in the generalization of NS data by different normally developed arable and grassland soil species or their groups. But for much more, we use the same matrix as well in the characterizing of abnormally developed soils. In this case, the scalar positions are mentioned by letter B, which marks the belonging of soils into series of abnormal ones, that form (figuratively saying) a second layer (story) above the main matrix.

Table 8.2 WRB's reference soil groups (RSG) and qualifiers of studied soil species and their position in relation to normally developed soils' matrix scalars (see Figure 8.1)

Code of Soil Species	***RSG of WRB***	***Characterization of Soil Properties with WRB Qualifiers***	***Position on Matrix Scalar***[1]		***Soil Group***
			Hydr	***Lit-gen***	
Kh	*Leptosols*	*leptic lithic rendzic calcaric hyperhumic*	0–10	0–10	I
Kr	*Cambisols*	*leptic skeletic rendzic calcaric*	0–10	10–20	I
K	*Cambisols*	*calcaric hyperhumic*	10–20	10–20	I
Ko	*Cambisols*	*cambic endocalcaric mollic humic*	10–20	20–30	I
KI	*Luvisols*	*luvic humic loamic*	10–20	30–40	I
LP	*Retisols*	*glossic stagnic fragic umbric epidystric*	20–30	40–50	I
Lk	*Podzols*	*albic umbric epidystric endogleyic*	10–20	50–60	I
Khg	*Leptosols*	*gleyic leptic lithic rendzic calcaric*	30–40	0–10	I
Kg	*Cambisols*	*skeletic calcaric hyperhumic endogleyic*	30–40	10–20	I
Kog	*Cambisols*	*endocalcaric humic endogleyic loamic*	30–40	20–30	I
KIg	*Luvisols*	*luvic humic endogleyic loamic*	30–40	30–40	I
LPg	*Retisols*	*glossic stagnic umbric endogleyic*	30–40	40–50	I
Lkg	*Podzols*	*albic umbric epidystric endogleyic*	30–40	50–60	I
Gk	*Gleysols*	*epioxygleyic calcaric mollic skeletic*	40–50	10–20	I
Go	*Gleysols*	*epioxygleyic calcaric mollic endoskeletic*	40–50	20–30	I
G(o)	*Gleysols*	*epioxygleyic mollic eutric*	40–50	20–30	I
GI	*Gleysols*	*reductigleyic dystric luvic umbric*	40–50	30–40	I
LkG	*Podzols*	*epigleyic spodic albic dystric rustic*	40–50	50–60	I
Go1	*Gleysols*	*Histic*	50–60	20–30	II
GI1	*Gleysols*	*Histic*	50–60	30–55	II
M3	*Histosols*	*saprihistic rheic eutric*	60–90	20–40	III
M2	*Histosols*	*hemihistic rheic eutric*	60–90	40–50	III
AG	*Fluvisols*	*epigleyic*	B40–50	B20–30	IV[2]
AG1	*Fluvisols*	*histic*	B50–60	B30–50	IV
AM	*Histosols*	*fluvic*	B60–90	B30–50	IV
Gr	*Gleysols*	*salic epigleyic*	B40–50	B20–40	V
Gr1	*Gleysols*	*salic histic*	B50–60	B30–50	V
ArG	*Fluvisols*	*salic epigleyic protic*	B40–50	B20–30	V

[1]For location range of different soils by moisture (Hydr) and litho-genetical (Lit-gen) scalars see Figure 8.1; [2]by B (before scalar position) the belonging of soils into abnormally developed soil group is mentioned.

On the agriculturally used land, the normally developed soils of group I (ca 80%) are dominated (Table 8.1). In the northern part of Estonia for the best arable soils are typical leached (Ko) and eluviated (KI) automorphic loamy soils with more or lesser extent calcareous automorphic loamy rendzinas (K). Named soils are in most cases on agricultural landscapes associated with their gleyed analogs (Kog, KIg and Kg), which are also with high potential fertility, but need for their universal use the selective artificial drainage. In the southern part of Estonia for the best arable soils are the pseudopodzolic soils (LP) which is associated with automorphic sod-podzolic (Lk) and their both gleyed analogs.

On vast lowland agricultural landscapes (first of all in the western part of Estonia) for arable land to a large extent the different gley-soils that associated them different kinds of gleyed soils' species which are in use as croplands. It is as a rule that the gley-soils and distributed on the lower part of landscapes peaty and fen soils need profound drainage in their use for agricultural purposes.

By the long-term conventional tradition, the normally developed peaty and peat soils (groups II and III) are preferred for utilization as grasslands. As is seen from Table 8.1 the fen soils, especially those with well-decomposed peats, are regarded very highly as well in quality of arable lands. What is applicable to regularly inundated (abnormal) alluvial and coastal soils (groups IV and V), then practically all of them can be used for natural grasslands. Therefore, the share of abnormal soils is relatively modest among agriculturally used areas. In relation to tested agriculturally used territory, the abnormal mineral soils formed 7.7% but abnormal organic soils only 1.1%.

Table 8.3 Erosion-affected soils (group VI): Names by ESC, WRB's RSG and qualifiers, and position on matrix's scalars

Code[1]	*Name*	*RSG of WRB*	*Most Essential WRB Qualifiers*	*Position on Matrix's Scalars*[2]	
By Estonian Soil Classification				*Hydr*	*Lit-gen*
Ke	Slightly eroded K soils	*Cambisols*	*calcaric skeletic*	B10–20	B10–20
Koe	Slightly eroded Ko soils	*Cambisols*	*endocalcaric cambic*	B10–20	B20–30
KIe	Slightly eroded KI soils	*Luvisols*	*luvic endoeutric*	B10–20	B30–40
LPe	Slightly eroded LP soils	*Retisols*	*glossic/retic umbric fragic*	B10–20	B40–50
E2k	Moderately eroded calcareous soils	*Regosols*	*aric calcaric*	B0–10	B10–20
E3k	Severely eroded calcareous soils	*Regosols*	*aric protic calcaric*	B0–10	B10–20
E2o	Moderately eroded leached soils	*Regosols*	*aric endocalcaric eutric*	B0–10	B20–40
E2I	Moderately eroded podzolic soils	Regosols	aric dystric	B0–10	B40–60
KId	Slightly deluvial KI soils	*Luvisols*	*luvic colluvic endoeutric*	B15–25	B30–40
LPd	Slightly deluvial LP soils	*Retisols*	*glossic/retic umbric colluvic dystric*	B15–25	B40–50
D	Automorphic deluvial soils	*Cambi-Luvi-Retisols*	*colluvic relocatic transportic pachic*	B10–25	B30–50
Dg	Gleyed deluvial soils		*colluvic gleyic relocatic transportic pachic*	B30–40	B30–50

[1]For uneroded soils with codes K, Ko, KI and LP see Tables 8.1 and 8.2; [2]B before scalar positions indicates the belonging to synlithogenetic soils' group.

Most of the abnormal mineral soils are erosion-affected soils (Table 8.1). Presented on the basis of DS MNA–2 in Table 8.3 erosion-affected soils' association should be taken as one exception among a huge quantity of possible alternatives. Presented DS characterizes erosion-affected agricultural landscape soil cover of locality Valgjärve. By these data (by this DS) is not possible to characterize the composition of soil species on relation to certain erosion-affected area, but the DS is well suitable for characterization of found on this area separate soil species. In our work not only their NS and connected with this soil properties but as well the pedo-ecological background of their formation is analysed.

The distribution erosion-affected landscapes in Estonia and their soil cover taxonomical composition by soil species and varieties have been preliminary profoundly treated in different publications (Kask 1975, Kokk 1977, Kõlli et al. 2010b). By these work it may be concluded that water erosion has only a marked influence in the south-eastern part of Estonia on hilly end moraine areas. In erosion hazardous counties (Võru, Valga), 32–37% of arable soils have been affected by erosion.

PRESENTATION OF SOILS' NITROGEN STATUS DATA BY SOIL GROUPS

Normally Developed Mineral Soils

Characterization nitrogen status of arable soils' humipedons

Although the NT concentration variations of arable soils' humipedons by different sites are relatively high, it is clearly visible the NT concentrations regular changes in connection with soil pedo-ecological properties (Figure 8.2). Similar to automorphic soil, in the sequences from extremely calcareous (Kr) to non-calcareous (Lk) soils the NT concentrations as well in gleyed soils (from Kg to LPg) and in gley-soils (from Gk to LkG) are decreased. In a lesser extent, the mentioned above regularities are seen as well in the case of NT stock densities per humipedon layer. It should be mentioned that in connection with using the gleyed (i.e., endogleyic or moist) soils and gley-soils (i.e., epigleyic or wet soils) in agricultural purposes, they all have been preliminary artificially drained. Although in consequence of drainage the NT contents, as a rule, are decreased in all soils, it has been in soils to a remarkable extent and preserved pre-drainage existed differences between soil types. In general cases the ratio C:N in humipedons is increased from automorphic soils to wet gley-soils as well as from calcareous soils to non-calcareous ones.

Characterization of soils nitrogen status in relation to whole soil cover

Besides of humipedon's NS, i.e., NT contents (g kg^{-1} and Mg ha^{-1} or NT concentration and stock density) and NT ratio with OC (C:N), is also necessary to study in the level of soils species in the NS of subsoil and in some extent as well as the substrates of the SC. For characterization NS of whole SC the DB Pedon is used here. DB Pedon and contained in it DS PDN–3 for arable land and DS PDN–4 for grasslands NT density data were calculated on the basis of separate soil genetic horizons data. Therefore, the summarized by soil cover layers (HP, SS and PM) NS data as well as is calculated by them; SC and ML NT densities per area in addition to HP layer are by their issue is the integrated data. So, calculated on the basis of them NT concentration is rather the mean weighted (by mass) concentrations, which may be used only as a generalized index in comparative ecological analyzes. In any case, they are not presented in an actual case.

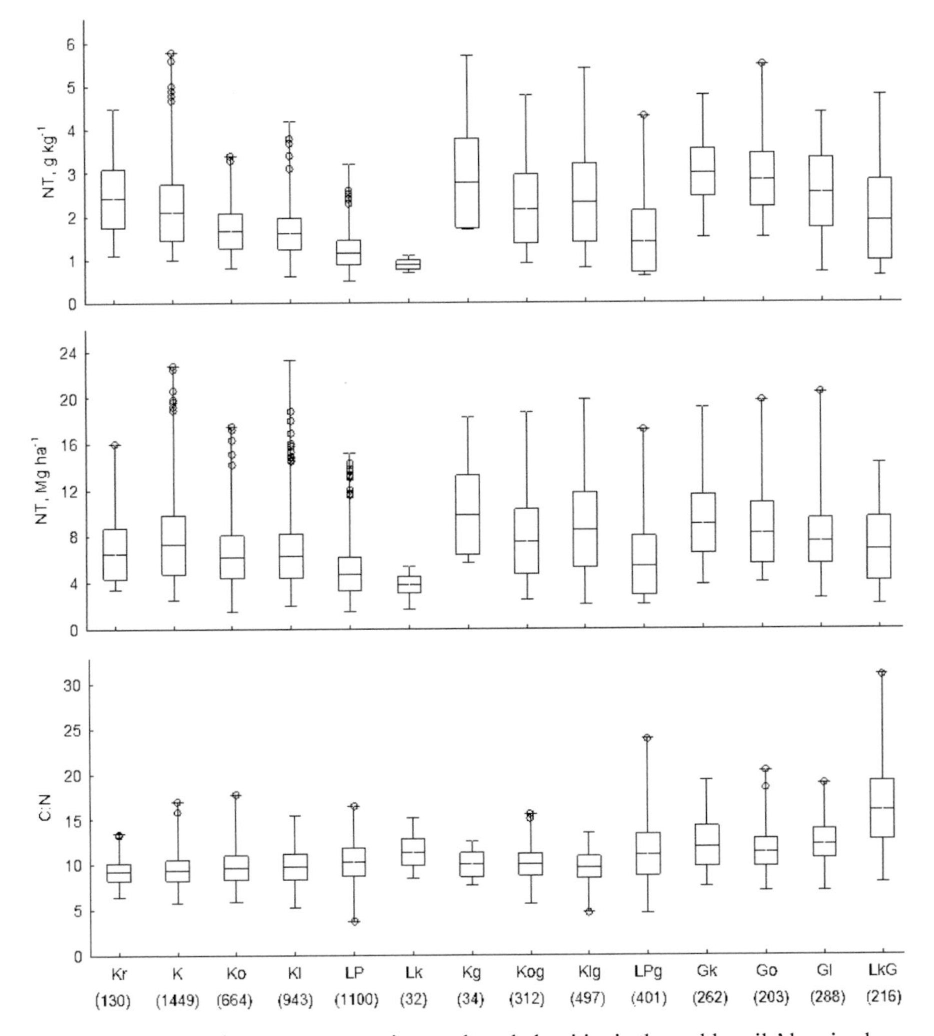

Figure 8.2 Total nitrogen concentrations and stock densities in the arable soils' humipedons were calculated on the basis of DS MNA–1 in the level of (for different) soil species. On the figure are presented: — Mean, □ Mean ± SD; I Min-Max and ○ Outliers. In brackets below the figure, a number of samples are given. For the soil names after their codes see Table 8.1.

As it is seen from Figure 8.3, the NT stock densities of presented arable soils' HP are in general very close to them of DS MNA–1 as presented in Figure 8.2. At the same time, the NT stocks in grassland soils exceed to certain extent stocks of arable soils HP layers. The NT stocks in SS depend first of all on the soil profile fabric. SS NT stocks are very modest in calcareous as well as endocalcareous gley-soils. But in soils with well-developed B and eluvial-illuvial horizons' sequence, the SS NT attaints to 4 and more Mg NT ha^{-1}. Belonging into substrate composition NT stocks depend on the case of normally developed mineral soils mainly on solum thickness and for being in a negative correlation with it.

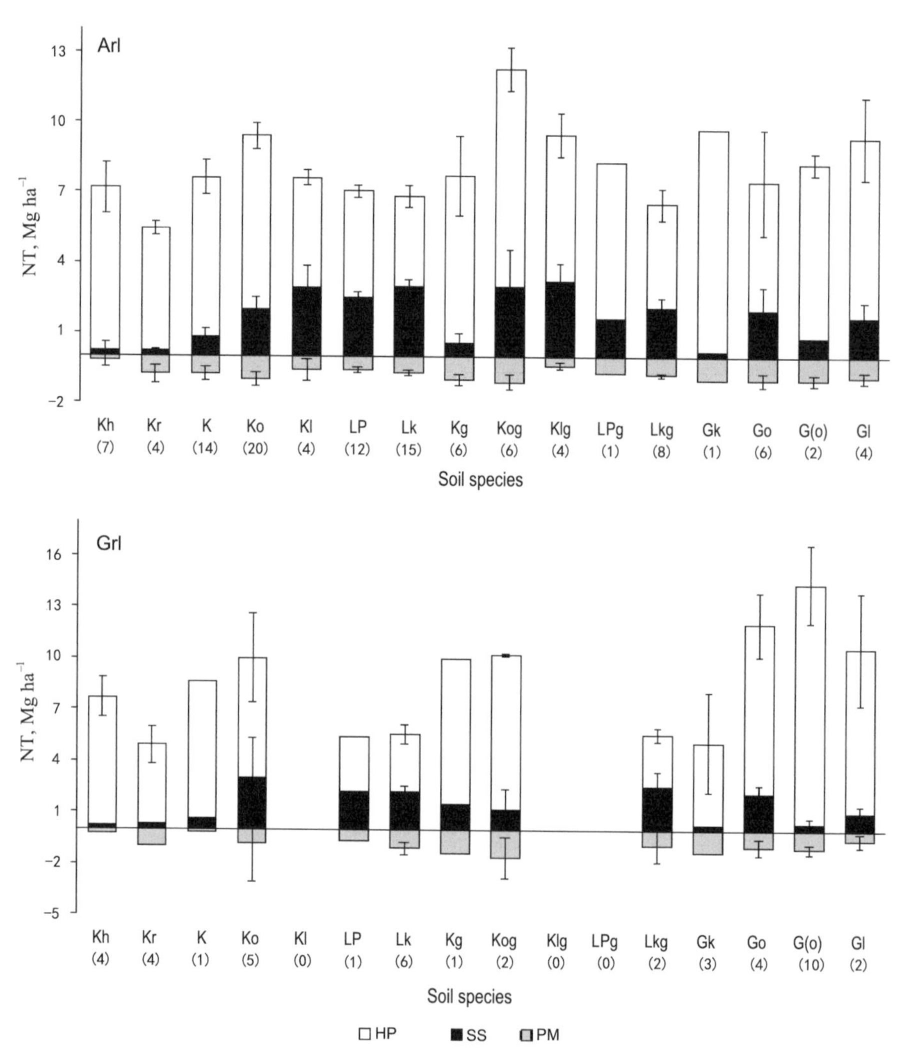

Figure 8.3 Total nitrogen stock densities in the different soil cover layers of arable and grassland soils calculated on the basis of DS Pedon–3 and Pedon–4 accordingly in the level of (for different) soil species.

Land use: Arl— arable and Grl— grasslands. Divisions on the bars: HP— humipedon, SS— subsoil and PM— parent material layers. On the figure, the Means ± SE are presented. In brackets below the figure, a number of research areas (RA) are given. For the soil names after their codes see Table 8.1.

Normally Developed Peaty and Peat Soils of Grasslands

The humipedons of peaty soils and whole profile of deep fen soils' ML are the richest in NT and OC soil covers of an agricultural landscape in comparison with all other agriculturally used soils. Among peaty soils, the richest in NT are formed on saturated gley-soils' peat horizons or peaty HP-s. At the same time, the SS and PM of peaty soils are very poor in NT. It is seen that account of these layers to the total ML NT stocks only approximately two Mg of NT is added. Of course, the SS and PM of deep fen soils are rich in NT, but in most cases, the NT stocks deposited deeper than 0.5 m may be accounted as buried NT stocks, which do not take part from biological cycling between soils and plant cover. The ratio C:N is as a rule larger in less saturated or more acidic soils.

Table 8.4 Total nitrogen (NT) concentrations, stock densities and C:N ratios in the different peaty and peat grasslands' soil cover layers calculated on the basis of DS Pedon–4

Soil Code[1]	*Matrix Scalar*		*Soil Layer*	*n*	*NS Indices*			
						NT Content		
	Hydr	*Lit-gen*			**h[2], cm**	**g kg^{-1}**	**Mg ha^{-1}**	*C:N*
			Humipedon	3	28	18.0	16.2	17.9
			Soil cover	3	40	6.8	17.1	17.4
Go1	55	26	Metric layer	3	100	1.29	17.5	17.1
			Subsoil	3	12	59 10^{-3}	0.9	8.9
			Parent material	3	60	3.6 10^{-3}	0.4	5.0
			Humipedon	3	25	11.9	9.5	27.4
			Soil cover	3	49	2.4	11.0	24.7
GI1	55	35	Metric layer	3	100	0.85	11.4	24.3
			Subsoil	3	24	40 10^{-3}	1.5	18.0
			Parent material	3	51	4.5 10^{-3}	0.4	12.5
			Humipedon	9	30	24.3	11.6	18.7
			Soil cover	9	50	24.0	19.2	18.7
M3	76	32	Metric layer[3]	9	100	24.1	38.7	19.2
			Subsoil	9	20	24.2	7.6	18.7
			Parent material[3]	9	50	24.4	19.5	20.0
			Humipedon	4	30	26.1	10.2	18.6
			Soil cover	4	50	28.4	15.6	18.1
M2	78	44	Metric layer[3]	6	100	22.0	27.2	22.7
			Subsoil	4	20	27.5	5.4	16.8
			Parent material[3]	6	50	17.9	11.6	29.1

[1]For soil names after their codes see Table 8.1; [2]Thickness of soil layer; [3]Calculated on the basis of DS FSP–7.

Abnormally Developed Alluvial and Coastal Soils of Grasslands

In the SC of alluvial and coastal soils are absent the eluvial and illuvial horizons, therefore their SS NT stocks are in most cases smaller than 2 Mg ha^{-1}. The exceptions from this aspect are alluvial fen and peaty coastal soils, which SS captured much more NT than rests, thanks to the highest NT concentration and/or to the biggest SS layer thickness. The NT concentrations of all alluvial and coastal soils' HP-s are relatively high, which is in concordance with their wet soil-forming conditions. Relatively low C:N ratio of all alluvial and coastal soils (groups IV and V) may be explained by the fact that in these abnormal soils the accumulated organic matter is much more than in normally developed soils mixed with mineral soil particles.

Table 8.5 Total nitrogen (NT) concentrations, stock densities and C:N ratios in the humipedon and subsoil layers of alluvial and coastal soil cover calculated on the basis of DS Pedon–4

Soil Code[1]	*Matrix Scalar*		*Soil Layer*	*n*	*NS Indices*			
	Hydr	*Lit-gen*			h^2, cm	*NT Content*		*C:N*
						g kg^{-1}	Mg ha^{-1}	
AG	B45	B40	Humipedon	3	30	4.63	12.1	8.2
			Soil cover	3	34	3.88	12.3	8.2
			Subsoil	3	4	0.36	0.2	8.2
AG1	B55	B38	Humipedon	4	23	10.66	8.9	14.4
			Soil cover	4	36	3.33	10.3	14.4
			Subsoil	4	13	0.62	1.4	14.4
AM	B68	B36	Humipedon	8	30	16.28	7.7	16.8
			Soil cover	8	50	16.92	13.3	16.8
			Subsoil	8	20	17.89	5.6	16.8
Gr	B45	B40	Humipedon	26	16	6.45	9.8	8.5
			Soil cover	26	42	2.79	11.7	8.2
			Subsoil	26	26	0.48	1.9	6.3
Gr1	B55	B45	Humipedon	5	14	20.24	6.0	11.4
			Soil cover	5	37	2.90	10.7	11.9
			Subsoil	5	23	1.22	4.7	12.1
ArG	B95	B30	Humipedon	5	3.8	14.90	4.0	12.7
			Soil cover	5	4.6	13.40	4.1	11.9
			Subsoil	5	0.8	5.26	0.1	10.0

[1]For soil names after their codes see Table 8.1; [2]Thickness of soil layer.

Erosion-affected Soils

The NS (NT concentration, NT stock density and C:N ratio) of erosion-affected soils is presented only for HP layers of arable soils (Figure 8.4). Because of erosion processes, the SC of erosion-affected areas is much more variegated in comparison with normally developed SC-s.

Although in HP of weakly eroded soils the NT concentrations are approximately at the same level with non-eroded soils, their difference becomes obvious by their NT stocks. In most cases, the lowest NT concentrations have moderately eroded soils but the highest deluvial soils. The mentioned above regularities described on the basis of NT concentration are proved by the data on NT stocks in arable soils HP layers. The influence of soil erosion is revealed clearly in NT stock densities. If non-eroded soils NT stock density is more than 4 Mg ha^{-1}, then in slightly and moderately eroded it is decreased to 1.4–3.9 Mg ha^{-1} but is increased in slightly deluvial (mentioned by d) soils to 5.5–7.3 and deluvial soils (D and Dg) to 12.2–14.5 Mg ha^{-1}.

On influenced by erosion landscape in a certain extent are presented as well the non-eroded automorphic soils (K, Ko, KI and LP), but as revealed, their NT concentrations and as well stocks densities are a little bit lower as compared with normally developed soil species (Figure 8.2). What pertains to ratio C:N, then it is mostly in the limits from 8.2 to 10.1 but being a little bit higher in eroded soils as well as more acid soils (LP).

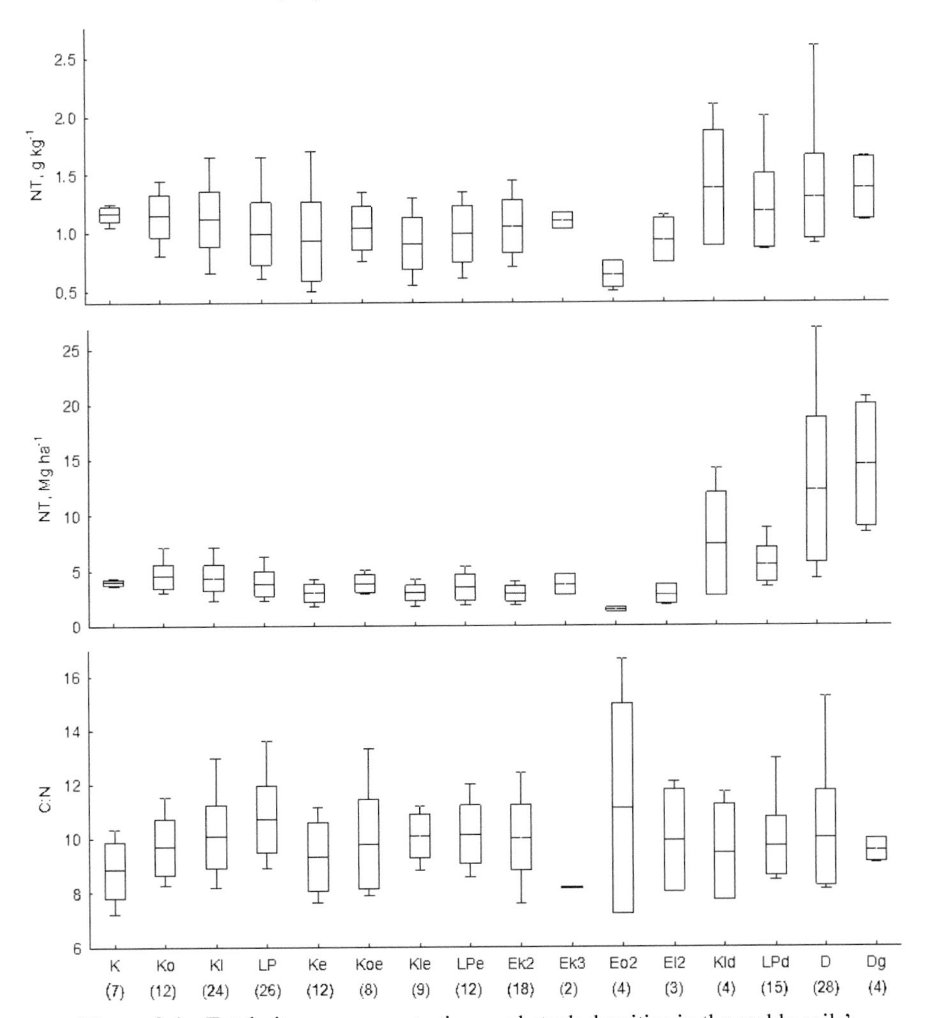

Figure 8.4 Total nitrogen concentrations and stock densities in the arable soils' humipedons on influenced by erosion landscape calculated on the basis of DS MNA–2 in the level of (for different) soil species.

On the figure are presented: — Mean, □ Mean ± SD; I Min-Max. In brackets below the figure, a number of samples are given. For the soil names after their codes see Table 8.3.

THE DEPENDENCE OF SOIL NITROGEN STATUS ON SOIL HUMUS STATUS AND ON SOIL MORPHOMETRIC, PHYSICAL AND CHEMICAL PROPERTIES

Ratio C:N in Soils and the Dependence of Humipedon Total Nitrogen Content on its Organic Carbon Content

The ratio C:N as pedo-ecological index of soil NS was calculated for all studied soils in both ways on the basis OC and NT concentration or also on the basis of their stocks. Ratio C:N in relation to HP of soil group I is presented in Figure 8.2 but of soil group VI on Figure. 8.4. The mean ratios are increased from automorphic to gley-soils being 9.2–11.4 in automorphic, 9.7–11.1 in gleyed and 11.0–16.0 in gley-soils. In the case of erosion-affected soils, any clear regularities in ratio C:N was not observed as the mean C:N ratios were in limits 9–10 with some exceptions.

In peaty humipedons, the C:N ratio is increased from 18 (eutric) to 27 (more acidic or dystric) (Table 8.4). In mineral substrate of peaty soils the C:N ratio is decreased considerably being in saturated soils approximately 5 and unsaturated ones 12. The underlaid peat SC substrate's C:N ratio may be very different depending on-site ecological conditions.

The ratios C:N of alluvial and coastal soils' layer are very similar each to other (Table 8.5). In the case of gley-soils (AG and Gr) this ratio is relatively low (between 8 and 9) as compared with normally developed gley-soils. In the presence of peaty or/and raw-humus materials in HP-s C:N ratio increases but stays always lower than it is in normally developed soils with comparable moisture conditions.

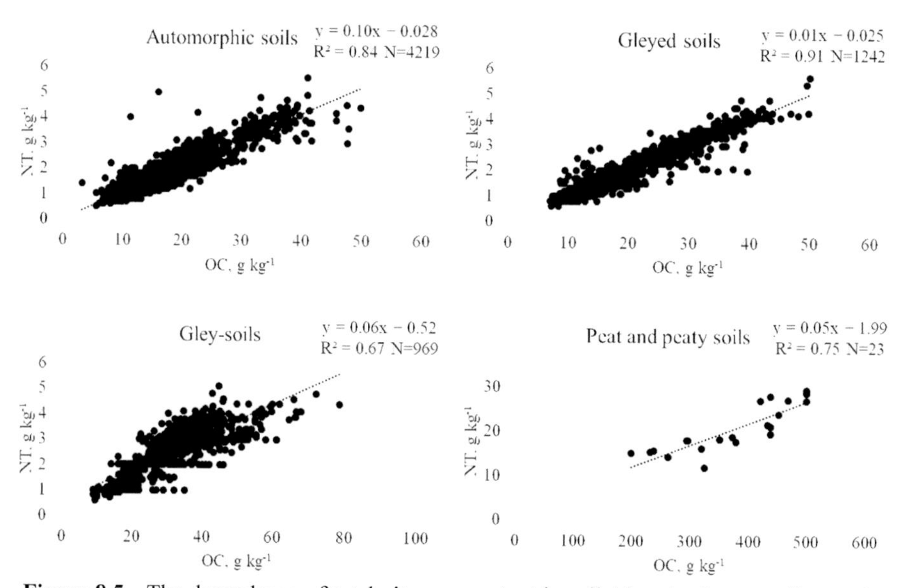

Figure 8.5 The dependence of total nitrogen content in soils' humipedon on soil organic carbon contents in pedo-ecologically different soil groups calculated on the basis of DS MNA–1, PDN–3 and PDN–4.

Thanks to high correlation between soil NT and OC contents it is possible to calculate NT contents (concentrations, stocks) on the basis of OC contents (Figure 8.5). It is proved by the fact that always in relation to whichever soil is accumulated much more data on OC as compared with data on NT contents. Generally, it is accepted that the most important argument determining soil NS is soil's humus status, i.e., managing soil nitrogen depends very much on its OC management. The NT contents in soil depend in analogs way with OC contents on soil moisture conditions. Therefore, deciding after given on Figure 8.5 formulas, it is meaningful in the calculation of soil NT contents indirectly to do it by different soil moisture groups (automorphic, gleyed, gley and peaty) of soils. The lowest correlation coefficient in the case of gley-soils is probably caused by their great differences in basic cation saturation stages.

Correlation of Nitrogen Status Indices with Main Arguments and with Some Soil Properties

As was concluded in the previous part, the NT contents have the highest correlation coefficients in relation to soil OC (Table 8.6). From the two pedo-ecological scalars the greatest influence on soil NT content has litho-genetic (Lit-Gen) scalar. The scalar of moisture conditions (Hydr) is an essential argument in the forming of C:N ratio. NT contents in the soil are correlated reliably also with the content of basic cations (BC), cation exchange capacity (CEC) and base saturation stage (BSS). NT contents have a reliable correlation with soil texture (calculated on the basis of physical clay), soil active acidity (pH) and thickness of the HP layer. By Hassink (1997) the capacity of soil to preserve nitrogen in soil depends on the texture of the soil, whilst the finest particles (silt and clay) of soil sequestrate much more nitrogen as compared with coarse fractions.

Demonstration and Analysis of the Background on which the Soil Nitrogen Status of Different Soil Species has been Formed

As the soil NS is determined in great extent by soil type and its properties is important to follow them in most detail level. To follow the NS and soil properties concordance is essential in both cases (i) in the presentation of basic experimental results (Figures 8.2–8.4 and Tables 8.4–8.5) and (ii) in the generalization of received information (given in next part). In connection with using a pedocentric research principle in actual work (see part 2.2), it should be emphasized that for the main taxonomical units of ESC in treating and analyzing NS results is the 'soil species' as acknowledged for practical use soil taxon. Therefore, the background soil properties in relation to NS are given mainly in the level of soil species (Tables 8.7 and 8.8). For the main exception among the group I soils are the sod-podzolic (Lk) and gleyed sod-podzolic (Lkg) soils, which have been not divided here by the podzolization stage (Table 8.7). On the soil species level, soil properties characterization is given only for the dominating in Estonian agricultural landscape soil species. Presented in both tables HP texture indicated to the possibly dominating soil variety as a sub-taxon of soil species.

Table 8.6 Correlation of soil nitrogen status (NS) indices (total nitrogen (NT) content and ratio C:N) with pedo-ecological characteristics of soils' humipedon (HP) layer

NS Indices	*DS*[2], *n*	*Soil Group*	*Scalar*		*Depth of HP, cm*	pH_{KCl}	*OC*	*EA*[3]	*HA*[3]	*BC*[3]	*CEC*[3]	*BSS*[3]	*PC*[3]	*Slope Gradient*
			Hydr	*Lit-gen*										
NT, g kg^{-1}	MNA–1,		0.46	–0.69*[4]	–0.59*	0.63*	0.93*	–	–	0.78*	0.81*	0.63*	0.85*	
NT, Mg ha^{-1}	(Arl, HP)	I	0.51	–0.58*	–0.31	0.52	0.84*	–	–	0.65*	0.69*	0.59*	0.69*	
C:N [C][1]	n = 14		0.78*	0.48	0.12	–0.52	0.32	–	–	–0.40	–0.35	0.62*	–0.10	
NT, g kg^{-1}	PDN–3,		–0.20	–0.43	–	0.23	0.71*	–0.40	0.07	0.53*	0.66*	0.23	–	
NT, Mg ha^{-1}	(Arl, HP)	I	0.28	–0.31	–	0.27	0.76*	–0.22	0.05	0.54*	0.50*	0.13	–	
C:N [C]	n = 17		0.28	0.16	–	–0.35	0.41	0.06	0.04	–0.11	–0.14	–0.09	–	
NT, g kg^{-1}	PDN–4,		0.39	0.02	–	0.18	0.48	0.11	0.19	0.41	0.49	–0.07	–	
NT, Mg ha^{-1}	(Grl, HP)	I	0.25	–0.40	–	0.37	0.85*	0.04	0.09	0.20	0.57*	0.24		
C:N [C]	n = 13		0.24	0.21	–	–0.47	0.10	0.10	0.41	–0.14	–0.10	–0.58*	–	
NT, g kg^{-1}	MNA–2,		–	–	0.09	0.07	–0.88	–	–	0.01	0.01	0.03	0.36	–0.74*
NT, Mg ha^{-1}	(Arl, HP)	VI	–	–	–0.08	–0.14	0.85*	–	–	–0.22	–0.22	–0.16	0.11	–0.73*
C:N [C]	n = 16		–	–	–0.28	–0.53*	–0.09	–	–	–0.53*	–0.53	–0.59*	–0.27	0.28

[1]C:N [C]— means that ratio C:N was calculated on the basis of soil organic carbon (OC) and NT concentration;

[2]The data sets are described in part 3.2;

[3]Agrochemical characteristics: EA— exchangeable acidity, HA— hydrolytic acidity, BC– basic cations, CEC— cation exchange capacity, BSS— stage of base saturation and PC—content of physical clay;

[4]*means reliable correlation.

Table 8.7 Pedo-ecological background of most studied soil species as excerpt from DB Pedon (mean ± SE)

Soil[1]	*N*	*Soil Group*	*HP*[2] *Texture*	*Land Use*	*Scalar*[3]		pH_{KCl}		*Depth of HP*, cm	*PC*[4], %	*HA*	*BC*	*CEC*	*BSS*, %	*Soil Quality*[5]
					Hydr	*Lit-gen*	*HP*	*PM*			kmol ha^{-1}				
Kh	11	I	$r_2$ls	Arl/Grl	08	06	6.7	7.1	23±1	29±2	38±6	636±43	674±46	95±1	30±1
Kr	8	I	$r_3$ls	Arl/Grl	07	16	6.8	6.9	18±2	24±3	20±3	627±98	647±99	97±1	33±4
K	14	I	$r_2$ls	Arl	13	16	6.8	6.9	27±1	14±2	31±3	747±83	777±84	94±2	54±1
Ko	20	I	ls	Arl	15	25	6.7	6.9	27±1	26±2	66±4	735±39	801±41	92±1	58±1
KI	4	I	ls	Arl	15	35	5.7	7.0	27±2	18±2	79±16	609±179	688±170	86±4	56±2
LP	12	I	sl	Arl	25	45	6.0	5.8	26±2	19±2	105±10	446±66	551±68	79±2	45±2
Lk	15	I	l, sl	Arl	15	62	6.3	5.7	23±2	13±2	92±12	439±75	531±78	78±4	37±3
Kg	6	I	$r_2$ls	Arl	35	16	6.5	6.8	27±4	27±3	40±3	853±145	893±148	95±1	58±4
Kog	6	I	ls	Arl	35	25	6.5	6.7	34±4	34±3	95±17	863±94	958±108	90±1	63±4
KIg	4	I	ls	Arl	35	35	6.3	7.4	28±3	25±3	96±25	479±124	575±148	83±1	57±3
Lkg	6	I	l	Arl	41	41	6.1	6.5	29±3	16±5	131±21	489±78	620±89	79±4	48±6
Gk	3	I	$r_2$ls	Grl	47	17	6.3	6.4	25±5	28±4	69±13	905±235	974±223	91±4	45±6
Go	7	I	ls	Grl/Arl	45	25	6.4	6.9	23±2	19±2	156±36	613±199	768±227	78±3	49±2
G(o)	10	I	ls, sl	Grl	45	26	5.8	6.5	29±2	28±4	157±29	753±100	910±119	83±3	52±2
GI	6	I	ls, sl	Arl/Grl	45	35	6.1	6.5	25±2	18±4	201±50	718±497	919±235	77±4	50±3
LkG	2	I	l	Grl	45	55	6.1	4.5	23±0	14±2	142±61	306±115	447±175	69±1	34±4
Go1	3	II	t3	Grl	55	26	5.9	6.3	28±3	peat	33±5	116±22	149±28	78±4	46±4
GI1	3	II	t2	Grl	55	35	4.9	5.5	25±4	peat	32±4	72±12	104±20	69±3	38±3
M3	9	III	t3	Grl	76	32	5.6	6.1	30	peat	189±16	594±48	778±49	72±4	53±5
M2	4	III	t2	Grl	78	44	4.9	5.5	30	peat	156±33	341±70	497±78	67±6	44±4
AG	3	IV	ls, sl	Grl	45	40	6.3	6.5	30±5	22±8	200±52	842±323	1043±373	79±2	39±4
AM	8	IV	t2	Grl	68	36	6.0	6.0	30	peat	102±14	349±30	451±41	78±2	37±2
Gr	10	V	sl, ls	Grl	45	40	6.0	–	17±2	–	43±11	555±218	598±217	67±6	–

[1]For the soil names after their codes see Table 8.1; [2]For symbols of dominating soil texture see pages 167–168; [3]For the position of soil species among other soils see Figure 8.1; [4]Physical clay (particles with ø <0.01 mm); [5]Soil quality in points taken according to ESC instructions (Astover et al. 2013).

The position of soils on the soil matrix (Figure 8.1) is given for soil group I–V (Table 8.7). By the soil species position in relation to moisture and litho-genetic scalars of the soil matrix enables a better understanding of the differences between various soil species and the tendency of properties changing on the background of soil matrix.

An essential factor in the formation of soil NS is land use. In connection with this, Table 8.7 is indicated dominating land use by different soils, whereas all presented in Table 8.8 soils characterize only erosion-affected soils. Different from Table 8.7 data is Table 8.8 where the data on erosion-affected soils' OC densities and slope gradients are presented, as these indices have a high correlation with influenced by erosion soils (Table 8.6).

Table 8.8 Pedo-ecological background of erosion-affected arable soils' humipedons (group VI) calculated on the basis of DS MNA–2

Soil[1]	*N*	*HP Texture*[2]	*HP,* cm	*OC,* Mg ha^{-1}	*Slope Gradient*	pH_{KCl}	*CEC,* mmol kg^{-1}	*BSS,* %	*Soil Quality*[3]
K	7	$r_2$ls	22.6	35	5.6	7.1	43.9	99.1	48
Ko	12	ls	24.4	43	5.8	7.0	31.0	98.1	52
KI	24	ls	24.3	44	6.0	6.5	16.9	93.5	50
LP	26	sl	24.8	41	7.3	6.0	10.3	84.5	45
Ke	12	$r_2$ls	21.9	28	8.3	7.2	46.7	99.3	39
Koe	8	ls	22.3	37	8.5	6.7	22.3	93.9	46
KIe	9	ls	19.8	29	7.9	6.4	22.4	93.4	45
LPe	12	sl	21.2	35	9.0	6.2	15.0	89.8	38
E2k	18	$r_2$ls	18.2	29	7.8	7.1	47.9	99.3	35
E3k	2	$r_3$ls	21.5	30	12.0	7.1	48.6	99.3	30
E2o	4	ls	13.5	16	9.5	5.6	6.9	80.2	37
E2I	3	ls	17.7	27	7.7	6.7	26.7	96.6	38
KId	4	ls	33.3	71	5.5	6.7	23.9	96.0	52
LPd	15	sl	29.9	53	7.7	5.9	14.7	81.2	48
D	28	sl	60.4	125	6.6	6.3	16.7	90.0	50
Dg	4	ls, sl	72.3	139	4.5	6.3	14.9	90.6	47

[1]For the soil names after their codes see Table 8.3; [2]For symbols of dominating soil texture see pages 167–168; [3]Soil quality in points taken according to ESC instructions (Astover et al. 2013).

GENERALIZED DATA ABOUT NITROGEN STATUS OF SOILS PRESENTED ON ESTONIAN AGRICULTURAL LANDSCAPE ON THE BACKGROUND OF THE SOIL MATRIX

Normally Developed Soils

The most important part of whatever soil is its HP layer, which in dominating cases is equal to soil humus horizon and/or plowing layer. Therefore, the generalized NS indices (NT concentration, NT stock density and C:N ratio) of agriculturally used soils are presented first of all in relation to this layer (Figure 8.8). The presented data on NS of Estonian agriculturally used soils should be taken as benchmark level data, which characterize the soil status at the last half-century, i.e., at the period when most of the data used here were collected. These generalized data are by their issue mean weighted by soil mass and taken from different sites data, whereas generalized in Figure 8.6 soils form approximately 91% from whole agriculturally used soils. As was seen from initial data (Figures 8.2 and 8.3), the NS of each site has as well as certain individual peculiarities. Therefore, each concrete site's NS indices may differ considerably from the mean levels of NS indices presented in Figure 8.6. But in many cases, the benchmark level data enable in giving a comparable estimation to the NS of the studied individual site, i.e., its difference from dominating status.

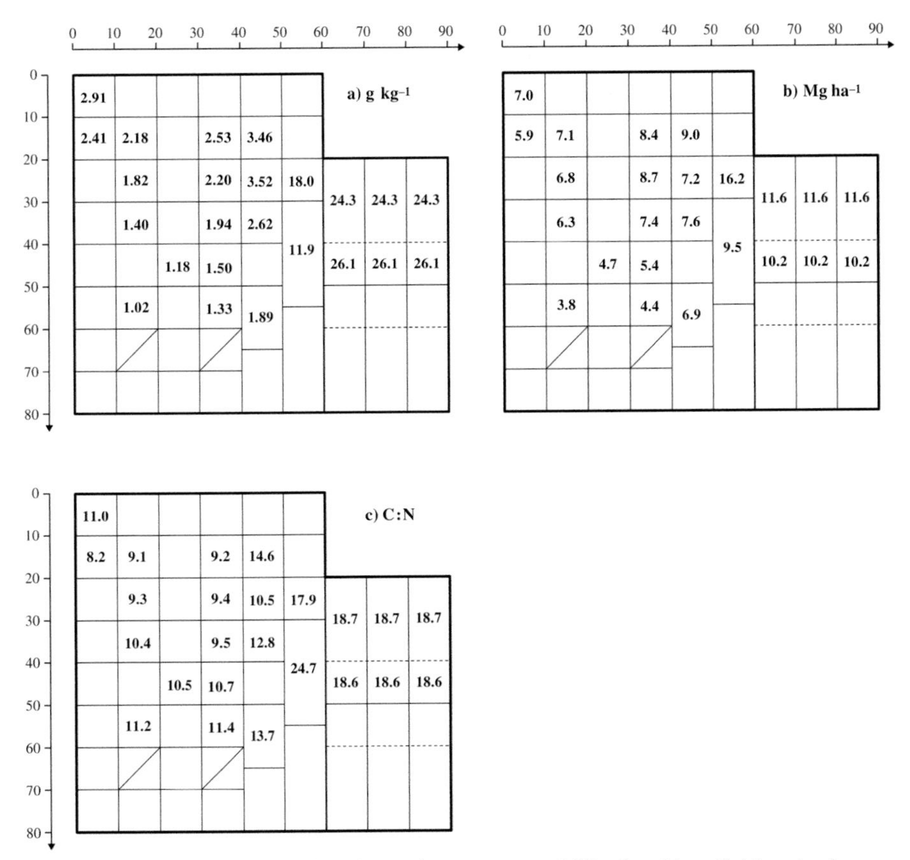

Figure 8.6 Generalized data about nitrogen status (NS) of arable soils' humipedons
a) Total nitrogen (NT) concentration (g kg^{-1}), b) NT stock density (Mg ha^{-1}) and c) NT relationship with OC (ratio C:N). For the soil codes and position on soil matrix see Figure 8.1, but for soil names after their codes see Table 8.1.

With the group I soils are treated on the same matrix as well as belonging to normally developed peaty (group II) and peat (group III) soils (Table 8.1). From them, the peaty soils, which are the transitional soils between mineral and organic soils, are very sensitive to arable land use. In the case of their tillage instead of peaty HP, the topsoil with raw-humus is formed, which means that instead of peaty soils the gley-soils may be formed. So, the given in Table 8.4 indices are characterized as used for grasslands. In the case of grassland, the described above drastic changes are evited.

One aim of generalizing NS data (Figure 8.6) is to demonstrate an essential tendency in changes of NT concentrations and stock's densities, and in C:N ratios in connection with changes in SC pedo-ecological background. For such kind of analysis, the 'mother' matrix (Figure 8.1) with an explanation on moisture conditions and litho-genetic scalars may be used. Besides pedo-ecological characterization on the basis of scalar positions is possible to see here the matching of soil names given by ESC and WRB (Table 8.2). Described above analysis is also expedient to look at the different soils' background data presented in Table 8.7. Besides soils names and positions on matrix, here the information about soils' HP thickness, texture, acidity and some agrochemical properties is given.

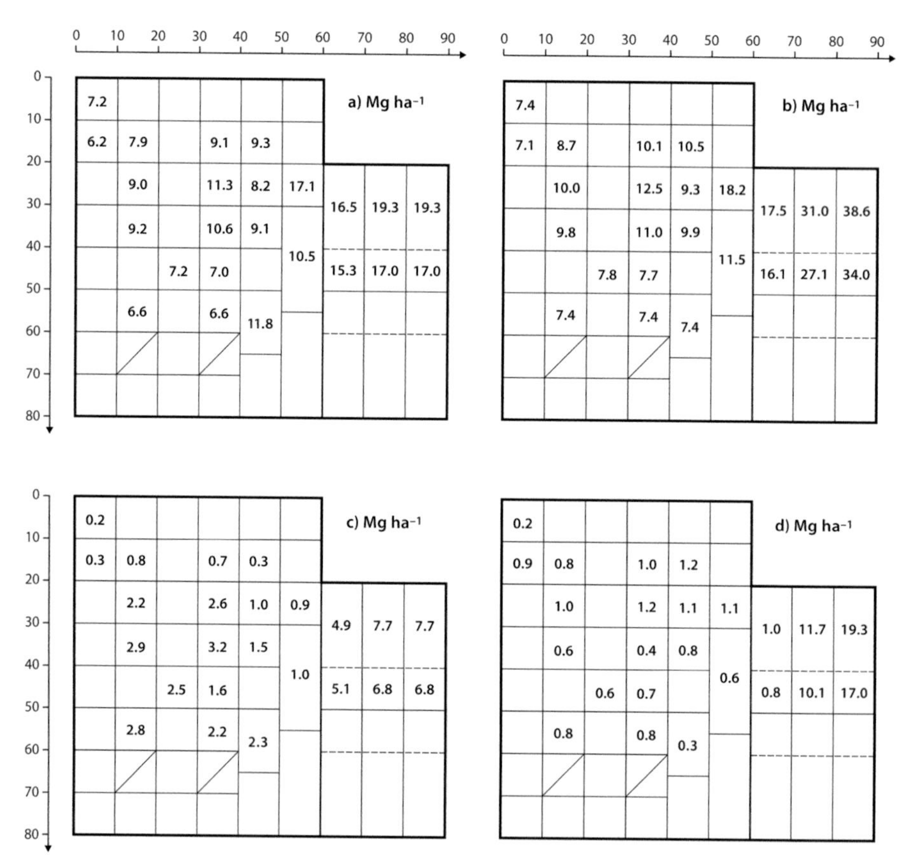

Figure 8.7 Mean weighted stock densities (in Mg ha^{-1}) of total nitrogen in agriculturally used lands' soil cover, metric layer and their sublayers

a) in soil cover or solum, b) in a metric layer of soil, c) in subsoil and d) in parent material or substrate. For the soil codes and position on soil matrix see Figure 8.1, but for soil names after their codes see Table 8.1.

By the thumb rule, the NT concentrations in mineral arable soils HP are increased from automorphic to permanently wet mineral soils from 1.0 to 3.5 g kg^{-1}. Incomparably higher are NT concentrations in peat and peaty soils, which are as the transitional soils between mineral and peat soils and are located in relation to moisture scalar on positions from 50 to 60 (Figure 8.6a). It is seen that NT concentrations in mineral soils are decreased from calcareous (eutric) to more acidic (dystric) soils. The HP's NT stocks changing tendencies on the background soil matrixes are analogic to their NT concentrations being in mineral soils HP in limits from 3.8 to 9.0 Mg ha^{-1} but much higher in peat soils (Figure 8.6b). Of course the ratio C:N is most wider in peat soils, decreasing step by step to direction of gley-, gleyed and automorphic calcareous soils (Figure 8.7).

In relation to normally developed soils by us are generalized NT stocks of SS and PM layers (Figure 8.7). SC's NT stocks contain besides sequestrated into HP NT stocks (Figure 8.7a) additionally these amounts of NT, which are sequestrated into SS layer (Figure 8.7c). The NT stocks of mineral soils' SS (among them of peaty soils) are in limits from 0.2 to 3.2 Mg ha^{-1} (Figure 8.7c). These data demonstrate that the biggest NT stocks have been sequestrated in a greater extent in the eluviated (podzolized) soils' SS, where the well-developed illuvial (B) horizons or sequences of eluvial-illuvial (E-B) horizons are found.

Although ML of soil contains (especially in Estonian pedo-ecological conditions) NT amounts, which do not participate in dominating cases in biological turnover, maybe ML NT stocks have taken as a good etalon in the comparable analysis of NS with alternative regions of the globe. The given in Figure 8.7b data have coincided relatively well with the data Batjes (2002) in relation to Cambisols, Luvisols, Podzoluvisols, Gleysols and Histosols. As was seen from Figure 8.7d, in dominating cases the amounts of the PM are lower from the SS ones.

Alluvial and Coastal Soils

The alluvial soils, which form from agriculturally used soils ca 2.3% are exploited mainly as grasslands. The alluvial soils belong to the landscapes with alluvial plains' localities (Arold 2005) and are distributed around the whole territory forming approximately 1–3% of typical agro-district's territory (Kokk and Rooma, 1974). As natural grasslands utilized as well the coastal soils, which are distributed on localities of marine plains. In Table 8.5, the main NS indices for HP, SC and SS layers of alluvial and coastal soils are presented. For alluvial and coastal soils is characteristic thin layered fabric, which is better visible in the case of sandy texture. Their HP thickness is exceptionally thin of coastal soils, but also of alluvial soils. But these thin layers are rich in both OC and coupled with it NT. Therefore, if all of alluvial and Gr of coastal soils' NT stock densities are in general lines very similar to normally developed gley-soils, then NT stocks of Gr1 and ArG are very low (<6 Mg ha^{-1}). In connection with soils of groups IV and V forming conditions, the variability of their NS (as humus status) is from site to site and is very variegated, therefore soil type-specific to NS is not clearly visible. Essential peculiarity connected with soil-forming conditions is as well as their SS and PM richness in NT and OC.

Data on alluvial and coastal soils' (as abnormal soils') NS have been generalized on the background of the abnormal soil matrix (Figure 8.8). The moisture conditions scalar (on horizontal bar) of this matrix (Figure 8.8a) match with the matrix of normally developed soils (Figure 8.1). The alluvial soils are located in the row 'A' (as mentioned on the vertical bar of the matrix), but the coastal soils are in rows 'r' and 'Ar' (Figure 8.8a). The possible position of these IV and V soil groups is shown in Table 8.2, but the mean studied DS (Pedon–4) soils' positions is in Table 8.5. The letter B before the scalar position should be taken as the indication of the second matrix layer over the normally developed soil matrix (Figure 8.1).

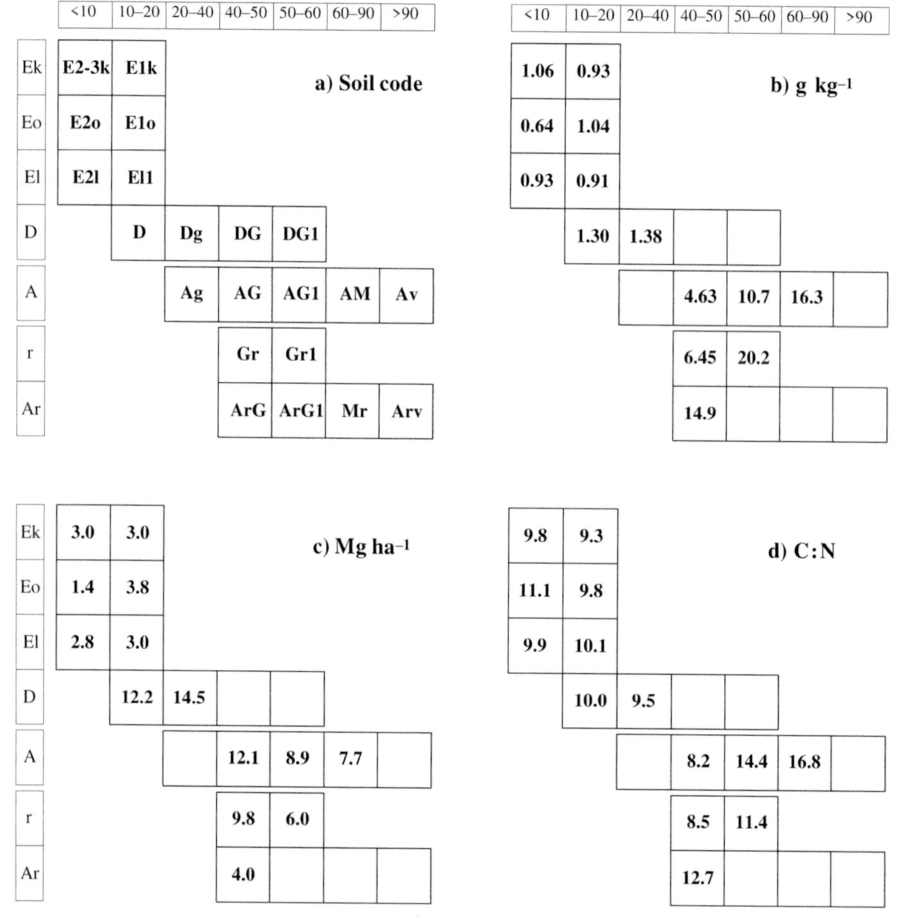

Figure 8.8 Mean total nitrogen (NT) concentration (g kg^{-1}) and NT stock density (Mg ha^{-1}) in humipedons of abnormally developed soils and the relationship of NT with OC (ratio C:N) by different soil types

a) codes of soil species, b) NT concentrations, c) NT stocks, and d) C:N ratio. For the soil names after their codes, see Table 8.3.

If Figure 8.8 has been characterized only by the NS of these soils' HPs, then Table 8.5 has presented some additional information about alluvial and coastal soils' SS and therefore about whole SC. Only dominating soil species of alluvial (AG and AM) and coastal soils (Gr) are given the characterizing of their pedo-ecological background data as the DS of studying groups IV and V were relatively small. The presented in this part soils' NS indices do not present the distributed in whole Estonian territory alluvial and coastal soil but only as representing the DS Pedon–4. However, they characterize well the high diversity of agricultural landscapes' NS.

Erosion-affected Soils

Erosion-affected soils as abnormally developed soils form a total 6.0% from agriculturally used Estonian soil cover (Table 8.1). The share of erosion-affected soils is high in Estonian South-eastern hilly end moraine landscapes (Kokk 1977, Kokk and Rooma 1978, Reintam et al. 2000). In agricultural landscapes, the erosion-affected soils are distributed to different extent depending on region soil-forming conditions. The core of erosion-affected soils association form two opposite groups of soils: (a) E– eroded (denuded, truncated and aric) and (b) D– deluvial or colluvial (thapto-, pachic, cumulinovic and transportic) soils. The fabric, formation and humus status (which is connected closely with NS) is treated in our previous work (Kõlli et al. 2010b). As is seen from Figure 8.4 and Table 8.3, the pedo-diversity of influenced by erosion areas is high, which is proved by the high number of soil species and their variants (mentioned by the letter d). High pedodiversity results with high diversity in NS of the landscape. The high pedodiversity is influenced by erosion areas that were mentioned by several other research (Olson et al. 2002).

The actual part generalized only the data on the NS of HP-s (Figure 8.8). The abnormally developed soil matrix (Figure 8.8a) should be taken as the second floor of normally developed soil matrix (Figure 8.1), whereas to follow their matching is possible after the soils' moisture and litho-genetic scalars given in Table 8.3. The eroded soils are located on rows Ek (calcareous), Eo (leached, endocalcic) and EI (podzolized) in relation to a vertical scalar of the matrix but the deluvial soils on row D (Figure 8.8a), corresponding to litho-genetic scalar's positions of normally developed soils (Figure 8.1) accordingly E soils on rows 10–60 and D soils 10–50. In an actual case under the E1k (Figure 8.8a), the soil Ke, under E1o the Koe and KIe, and under E1I the LPe (see Tables 8.3 and 8.8) are given.

In most cases, the mean NT concentrations of eroded soil species HP are in limits 0.6–1.6 g kg^{-1} but of deluvial (colluvial) soils' HP in 1.3–1.4 g kg^{-1}. There are seen clear differences in NT stock densities between weakly (3–4 Mg ha^{-1}) and moderately eroded (1.4–3.0 Mg ha^{-1}) soils' HP. Essentially high are the NT stock densities (12–15 Mg ha^{-1}) of deluvial (D and Dg) soils. It is interesting that there are not reliable differences between influenced by erosion soil species in C:N ratio. The ratio C:N is generally in limits from 9.3 to 11.1.

The affected by erosion agricultural landscapes' soil mantle on hilly areas is always much more variegated as compared with dominating flatty areas by soil species number, HP thicknesses, soil texture and agrochemical properties. Depending on erosion influence, soil humus status, texture and HP thickness, the soil quality is as well as very variegated (from 30 points to 52 and more). The most essential soil data characterizing pedo-ecological background of erosion-affected soils are given in Table 8.8.

ESSENTIAL CONCLUDING REMARKS UPON DETERMINATION OF AGRICULTURAL SOILS TOTAL NITROGEN STATUS

The NS of soil is similar to soil humus status with its different indices (NT concentrations and NT stock densities by different soil layers according to C:N ratios) and is the most important driving (influencing) factor that determines the soil functioning potentiality (productivity) and its suitability for plant cover. The comparative analysis of NS on the background of soils' pedo-ecological conditions matrixes revealed that the NT content and ratio with OC depend on soil properties and therefore the soil NS is a soil type-specific feature.

Having NT densities (Mg NT ha^{-1}) data for all dominating soil species by their HP, SS and PM layers is possible to calculate NT pools to a certain land massive, region or whole Estonia, as it was done by the Batjes (2002, 2014) in relation to certain soil groups, European regions and worldwide level. It should be mentioned that the soil mapping units list of Estonian large scale 1:10,000; soil map is practically equal to ESC's soil species list. Soil type-specific approach to soil NS enables to elucidate essential differences between various landscape types.

Similarly to the pedodiversity of landscape, the diversity of it NS may be caused by soil texture variations (from sand to clay), mineralogical and chemical composition, calcareousness and acidity. The pattern of SC and its diversity are induced by the geodiversity and hydrological conditions. The pedodiversity of the landscape with its NS is an abiotic basis for the formation of optimal (specific to soil type) biodiversity. For a better understanding of the mutual influences of SC and plant cover, the feedback influences of their main components (soil and plant) functioning should be studied at the ecosystem level and on typical-to-region soil types and management conditions. The NS of the landscape or NT throughout the flux of SC is tightly connected with plant cover composition, productivity and diversity.

In the formation of soil species NS are clearly seen regional peculiarities, caused by differences in soil properties and functioning. In the Estonian agricultural landscape, the regional peculiarities are more clearly followed in the case of peudopodzolic soils' and some species of gley-soils (Kokk and Rooma 1978, Teras and Rooma 1985, Rooma 1987). In the case of pseudopodzolic soils (LP and LPg), as transitional soil between calcareous and non-calcareous soils, the regional differences in NS are caused by the calcareousness of the parent material and topsoil texture. In the case of gley-soils (mainly of G(o) and GI soils) for the driving force of regional differences are moisture conditions or trophic stage of soil water.

For a good ecological indicator in characterizing outlines of ecosystem functioning seems to be humus cover type (pro humus form). The humus cover (or humipedon) types are tightly connected with soil NS, reflecting well agricultural ecosystems' productivity and suitability of soils to different crops (Kõlli 2018). The comparable analysis of arable soils' NS indices distribution (Figure 8.6) with the distribution of humus cover types on the same soil matrix reveals clear distribution regularities of NS as well by soils' humus cover types.

The NT availability in soil is an important influencing factor in SOM humification processes and therefore in OC sequestration in soil (Christopher and Lal 2007, Poeplau et al. 2018). The HP of Estonian automorphic calcareous arable soils in which the NT concentration is much higher as compared with more acid non-calcareous soils have also considerably narrower C:N ratio, accordingly 8.2–9.4 and 10.5–11.4 (Figure 8.6).

The ecologically sound matching of land use with SC's NS and humus status is of pivotal importance in reaching the sustainable functioning of agro-ecosystems and in assuring a good environmental status of the ambient area. The environmental protection ability of soils may be attained by the ecologically sound management of landscape by taking into account its nitrogen and humus status. With land-use change (from natural to arable and vice versa), the more drastic changes occur in the fabric and properties of HP (Baddeley et al. 2017), whereas the SS rests in an almost unchanged state (Köster and Kõlli 2013).

REFERENCES

Arinoushkina, E.V. 1970. Rukovodstvo po Himitšeskomu Analizu Potšv (Instruction for Chemical Analysis of Soil). University of Moscow, p. 487 (in Russian).

Arold, I. 2005. Eesti Maastikud (Estonian Landscapes). Tartu Ülikooli Kirjastus, Tartu, p. 453 (in Estonian).

Astover, A., Reintam, E., Leedu, E. and Kõlli, R. 2013. Muldade Väliuurimine (Field Survey of Soils). Eesti Loodusfoto, Tartu, p. 70 (in Estonian).

Baddeley, J.A., Edwards, A.C. and Watson, C.A. 2017. Changes in soil C and N stocks and C:N stoichiometry 21 years after land use change on an arable mineral topsoil. Geoderma. 303: 19–26.

Batjes, H. 2002. Carbon and nitrogen stocks in the soils of Central and Eastern Europe. Soil Use Management. 18: 324–329.

Batjes, N.H. 2014. Total carbon and nitrogen in the soils of the world. European Journal of Soil Science. 65: 4–21.

Bremner, J.M. 1960. Determination of nitrogen in soil by the Kjeldahl method. The Journal of Agricultural Science. 55: 11–33.

Chapman, L.Y., McNulty, S.G., Sun, G. and Zhang, Y. 2013. Net nitrogen mineralization in natural ecosystems across the conterminous US. International Journal of Geosciences. 4: 1300–1312.

Christopher, S.F. and Lal, R. 2007. Nitrogen management affects carbon sequestration in North American cropland soils. Critical Review Plant Science. 26: 45–64.

DeBusk, W.F., White, J.R. and Reddy, K.R. 2001. Carbon and nitrogen dynamics in wetland soils. pp. 27–53. *In*: Shaffer, M.J., Ma, L. and Hansen, S. (eds), Modelling Carbon and Nitrogen Dynamics for Soil Management. Lewis Publishers, Boca Raton, Florida.

Estonian Land Board. 2012. Vabariigi digitaalse suuremõõtkavalise mullastiku kaardi seletuskiri (Explanatory letter for the large scale digital soil map of Estonia). http://geoportaal.maaamet.ee/docs/muld/mullakaardi_seletuskiri.pdf, accessed 15.01.2019 (in Estonian).

Estonian Land Board. 2018. Haritava maa 2018. aasta turuülevaade (Market Survey of Arable Land 2018). https://geoportaal.maaamet.ee/est/Andmed-ja-kaardid/Maakatastri-andmed/Maakatastri-statistika-p506.html (in Estonian).

Fridland, V.M. 1982. Main Principles and Elements of Soil Classification Bases and Work Programme for Their Creation. Moscow, p. 149 (in Russian).

Garten, C.T. and Ashwood, T.L. 2002. Landscape level differences in soil carbon and nitrogen: Implications for soil carbon sequestration. Global Biogeochemical Cycles. 16(4): 1–14.

Hassink, J. 1997. The capacity of soils to preserve organic C and N by their association with clay and silt particles. Plant and Soil. 191: 77–87.

IUSS Working Group WRB. 2015. World Reference Base for Soil Resources 2014, update 2015. International soil classification system for naming soils and creating legends for soil maps. World Soil Resources Reports No 106. FAO, Rome.

Kachinsky, N.A. 1965. Fizica Pochvy (Soil Physics) Vol. I. University Press, Moscow, p. 323 (in Russian).

Kask, R. 1975. Eesti NSV Maafond ja Selle Põllumajanduslik Kvaliteet (Land Resources of Estonian SSR and its Agricultural Quality). Valgus, Tallinn, p. 358 (in Estonian).

Kauer, K., Köster T. and Kõlli, R. 2004. Chemical parameters of coastal grassland soils in Estonia. Agronomy Research. 2(2): 169–180.

Kokk, R. and Rooma, I. 1974. Agromullastikuline rajoneerimine (Dividing into agro-districts). ENSV Mullastik Arvudes. I: 16–30 (in Estonian).

Kokk, R. 1977. Kallakute ja erodeeritud muldade levikust Eesti NSV haritavatel maadel. (Spread of slopes and eroded soils on arable land in Estonia). Land Reclamation. II(8): 24–30 (in Estonian).

Kokk, R. and Rooma, I. 1978. Eesti NSV haritavate maade muldade mõningate keemiliste, füüsikalis-keemiliste ka füüsikaliste omaduste iseloomustus (Characterization on some chemical, physicochemical and physical properties of arable land soils of Estonian SSR). ENSV Mullastik Arvudes. II: 3–66 (in Estonian).

Kokk, R. and Rooma, I. 1983. Haritavad maad (Arable lands). ENSV Mullastik Arvudes. III: 3–26 (in Estonian).

Kokk, R. 1995. Muldade jaotumus ja omadused (Distribution and properties of soils). pp. 430–443. *In*: Raukas, A. (ed.), Estonia. Nature. Valgus, Tallinn (in Estonian).

Kõlli, R. and Ellermäe, O. 2001. Soils as basis of Estonian landscapes and their diversity. pp. 445–448. *In*: Mander, Ü., Printsmann, A. and Palang, H. (eds). Development of European Landscapes. Conference Proceedings, Volume II. Publicationes Instituti Geographici Universitatis Tartuensis, 92. IALE, Tartu.

Kõlli, R. and Ellermäe, O. 2003. Humus status of postlithogenic arable mineral soils. Agronomy Research. 1(2): 161–174.

Kõlli, R., Köster, T. and Kauer, K. 2007. Organic matter of Estonian grassland soils. Agronomy Research. 5(2): 109–122.

Kõlli, R., Astover, A., Noormets, M., Tõnutare, T. and Szajdak, L. 2009a. Histosol as an ecologically active constituent of peatland: A case study from Estonia. Plant and Soil. 315: 3–17.

Kõlli, R., Ellermäe, O., Köster, T., Lemetti, I., Asi, E. and Kauer, K. 2009b. Stocks of organic carbon in Estonian soils. Estonian Journal of Earth Sciences. 58(2): 95–108.

Kõlli, R. and Kanal, A. 2010. The management and protection of soil cover: An ecosystem approach. Forestry Studies. 53: 25–34.

Kõlli, R., Asi, E., Apuhtin, V., Kauer, K. and Szajdak, L. 2010a. Formation of the chemical composition of Histosols and histic soils in the forest lands of Estonia. Chemical Ecology. 26(4): 289–303.

Kõlli, R., Ellermäe, O., Kauer, K. and Köster, T. 2010b. Erosion-affected soils in the Estonian landscape: humus status, patterns and classification. Archives of Agronomy and Soil Science. 56(2): 149–164.

Kõlli, R., Köster, T., Kauer, K. and Lemetti, I. 2010c. Pedoecological regularities of organic carbon retention in Estonian mineral soils. International Journal of Geosciences. 1: 139–148.

Kõlli, R. and Tamm, I. 2012. Eesti parimate põllumuldade levik, rühmitamine ja huumusseisund (Distribution of best Estonian arable soils, their grouping and humus status). pp. 15–22. *In*: Metspalu, L., Viiralt, R. and Karp, K. (eds), Agronoomia. AS Rebellis, Saku. (in Estonian).

Kõlli, R. 2018. Influence of land use change on fabric of humus cover. Applied Soil Ecology. 123: 737–739.

Köster, T. and Kõlli, R. 2013. Interrelationships between soil cover and plant cover depending on land use. Estonian Journal of Earth Sciences. 62(2): 93–112.

Laasimer, L. 1965. Eesti NSV Taimkate (Vegetation of the Estonian SSR). Valgus, Tallinn, p. 397 (in Estonian).

Lehtveer, R. and Kokk, R. 1995. Põllumuldade Seire (Monitoring of Arable Soil). Eesti Maauuringud, Tallinn, p. 101 (in Estonian).

Mander, Ü., Palang, H. and Tammiksaar, E. 1995. Landscape changes in Estonia during the 20th century. Year-Book of EGS. 29: 73–97.

Marty, C., Houle, D., Gagnon, C. and Courchesne, F. 2017. The relationships of soil total nitrogen concentrations, pools and C:N ratios with climate, vegetation types and nitrate deposition in temperate and boreal forests of eastern Canada. Catena. 152: 163–172.

Olson, K.R., Gennadiyev, A.N., Jones, R.L. and Chernyanskii, S. 2002. Erosion patterns on cultivated and reforested hill slopes in Moscow regions, Russia. Soil Science Society of America Journal. 66: 193–201.

Orlov, D.S. and Grišina, L.A. 1981. Laboratory Course on Soil Chemistry. Moscow, p. 272 (in Russian).

Poeplau, C., Zopf, D., Greiner, B., Geerts, R., Korvaar, H., Thumm, U., et al. 2018. Why does mineral fertilization increase soil carbon stocks in temperate grasslands? Agriculture, Ecosystems and Environment. 265: 144–155.

Raukas, A. (ed.) 1995. Eesti. Loodus (Estonia. Nature). Valgus, Tallinn, pp. 606. (in Estonian).

Raukas, A. and Teedumäe, A. 1997. Geology and Mineral Resources of Estonia. Estonian Academy Publishers, Tallinn, p. 436.

Reintam, L., Rooma, I., Kull, A., Kitse, E. and Reintam, I. 2000. Soil vulnerability and degradation in Estonia. pp. 43–47. *In*: Batjes, N.H. (ed.), Soil Degradation Status and Vulnerability Assessment for Central and Eastern Europe. ISRIC Report 04.

Rooma, I. 1987. Kahkjate muldade levik ja omadused Eesti NSV-s (Distribution and properties of pseudopodzolic soils in Estonian SSR). ENSV Mullastik Arvudes. VI: 35–44 (in Estonian).

Shaffer, M.J. and Ma, L. 2001. Carbon and nitrogen dynamics in upland soils. pp. 11–29. *In*: Shaffer, M.J., Ma, L. and Hansen, S. (eds), Modelling Carbon and Nitrogen Dynamics for Soil Management. Lewis Publishers, Boca Raton, Florida.

Suuster, E., Ritz, C., Roostalu, H., Reintam, E., Kõlli, R. and Astover, A. 2011. Soil bulk density pedotransfer functions of the humus horizon in arable soils. Geoderma. 163: 74–82.

Suuster, E., Ritz, C., Roostalu, H., Reintam, E., Kõlli, R. and Astover, A. 2012. Modelling soil organic carbon concentration of mineral soils in arable land using legacy soil data. European Journal of Soil Science. 63: 351–359.

Szajdak, L.W. and Gaca, W. 2010. The influence of nitrogen on denitrification processes in soil under shelterbelt and adjoining cultivated field. pp. 225–235. *In*: Szajdak, L.W. and Karabanov, A.K. (eds), Physical, Chemical and Biological Processes in Soils. Prodruk, Poznań.

Teras, T. and Rooma, I. 1985. Kahkjate põllumuldade iseloomustus (Characterization of pseudopodzolic arable soils). ENSV Mullastik Arvudes. IV: 39–53 (in Estonian).

Tjurin, I.V. 1935. Comparative study of the methods for the determination of organic carbon in soils and water extracts from soils. pp. 139–158. *In*: Materials on genesis and geography of soils. ML Academy of Science USSR. (in Russian).

Vorobyova, L.A. 1998. Chimitsheskij Analiz Potchv (Chemical Analysis of Soils). Moscow University Press, Moscow, p. 260 (in Russian).

Xue, Z., Cheng, M. and An, S. 2013. Soil nitrogen distributions for different land uses and landscape positions in a small watershed on Loess Plateau. China Ecological Engineering. 60: 204–213.

Zanella, A., Jabiol, B., Ponge, J.F., Sartori, G., de Waal, R., Van Delft B., et al. 2011. A European morpho-functional classification of humus forms. Geoderma. 164: 138–145.

Zanella, A., Ponge, J.F., Gobat, J.M., Juilleret, J., Blouin, M., Aubert, M., et al. 2018. Humusica 1, Article 1: Essential bases–Vocabulary. Applied Soil Ecology. 122P1: 10–21.

Index